T0137334

NanoScience and Technology

The series NanoScience and Technology is focused on the fascinating nano-world, mesoscopic physics, analysis with atomic resolution, nano and quantum-effect devices, nanomechanics and atomic-scale processes. All the basic aspects and technology-oriented developments in this emerging discipline are covered by comprehensive and timely books. The series constitutes a survey of the relevant special topics, which are presented by leading experts in the field. These books will appeal to researchers, engineers, and advanced students.

More information about this series at http://www.springer.com/series/3705

Seongbo Shim · Youngsoo Shin

Physical Design and Mask Synthesis for Directed Self-Assembly Lithography

Seongbo Shim
Samsung Electronics
Hwasung
Korea (Republic of)

Youngsoo Shin
KAIST
Daejeon
Korea (Republic of)

ISSN 1434-4904 ISSN 2197-7127 (electronic)
NanoScience and Technology
ISBN 978-3-030-09455-3 ISBN 978-3-319-76294-4 (eBook)
https://doi.org/10.1007/978-3-319-76294-4

Printed on acid-free paper

This Springer imprint is published by the registered company Springer International Publishing AG part of Springer Nature
The registered company address is: Gewerbestrasse 11, 6330 Cham, Switzerland

For all those who have contributed to and will benefit from this field

Seongbo Shim and Youngsoo Shin

Preface

Beyond 7 nm technology node, traditional optical lithography is facing substantial challenges for printing fine features while maintaining a reasonable manufacturing cost. Alternative patterning solutions for the next generation lithography have actively been studied. They include extreme ultraviolet lithography, electron beam lithography, and multiple patterning lithography. Very high manufacturing cost or low throughput, however, has slowed down the adoption of these solutions. Another patterning solution named directed self-assembly lithograph (DSAL) is being considered practical, in particular for patterning contact holes and vias.

In DSAL, contacts (or vias) that are physically close are clustered and patterned together in two steps. (1) A large trench pattern that surrounds the contact cluster, called a guide pattern (GP), is first created on a wafer through optical lithography. (2) GP is then filled with block copolymers (BCPs), which are strings of two types of polymer; BCP in its nature is arranged due to forces between polymers and GP wall and one type of polymer is etched down to substrate, which leaves final contact holes.

GP serves as a bridge between optical lithography and DSA process; it is thus a key component of DSAL. Small variations on GP shape, which may arise during lithography process, can lead to incorrect contact patterning in DSA process. This poses a few challenges both in physical design and mask synthesis; this book is a result of our study to address such challenges for a few years. Contact topologies are limited due to the limitation on GP, which calls for careful consideration in design stages, e.g., during custom layout, placement, and routing. In mask synthesis, a set of GPs for a given contact layout can only be extracted through lengthy simulations; once GPs are obtained, they should be verified to see whether target contacts are formed, which is also a difficult problem.

This book is a concise introduction to DSAL technology and contains solutions to a number of circuit design and mask synthesis problems that we have identified. We hope that it serves as a good and popular reference for practical engineers and researchers both in lithography and CAD/design community.

Suwon, Gyeonggi-do, Korea
Daejeon, Korea
March 2018

Seongbo Shim
Youngsoo Shin

Contents

Acronyms

AF	Assist feature
ArF	Argon fluoride laser
BCP	Diblock copolymer
CD	Critical dimension
CPP	Contacted poly pitch
DOE	Diffraction optical element
DOF	Depth of focus
DP	Double patterning
DUV	Deep ultraviolet
DSA	Directed self-assembly
DSAL	Directed self-assembly lithography
EBL	Electron beam lithography
EPE	Edge placement error
EUVL	Extreme ultraviolet lithography
GDR	Gridded design rule
GP	Guide pattern
HVM	High volume manufacturing
KrF	Krypton fluoride laser
MCM	Maximum cardinality matching
MIS	Maximum independent set
MP	Multiple patterning
MP-DSAL	Directed self-assembly lithography with multiple patterning
NA	Numerical aperture
NGL	Next generation lithography
NIL	Nanoimprint lithography
OAI	Off-axis illumination
OPC	Optical proximity correction
PAG	Photo-acid generator
PCA	Principal component analysis
PVB	Process variation band

RET	Resolution enhancement technique
SADP	Self-aligned double patterning
SCFT	Self-consistent field theory
SEM	Scanning electron microscope
SRAF	Sub-resolution assist feature
SVM	Support vector machine
TP	Triple patterning
UV	Ultraviolet
193d	193 nm ArF dry lithography
193i	193 nm ArF immersion lithography

Chapter 1
Introduction

It is now well known that the scaling of devices is approaching fundamental as well as economic limit. This is mainly because traditional optical lithography is facing substantial challenges for printing fine features while maintaining a reasonable cost. Alternative patterning approaches for next generation lithography have been actively studied. Examples include extreme ultraviolet lithography (EUVL), electron beam lithography (EBL), and nanoimprint lithography (NIL). However, all of them involve high cost or low throughput, and so are not yet practical solution. Directed self-assembly lithography (DSAL), the focus of this book, is a practical solution and is believed to be the most promising technique for contact and via patterns in technology node of 7 nm and below.

In Sect. 1.1 of this chapter, optical lithography is briefly reviewed. It is then followed by the review of next generation lithography technologies including EUVL, EBL, and NIL in Sect. 1.2. DSAL is addressed in Sect. 1.3, and the overall structure of this book is presented in Sect. 1.4.

1.1 Optical Lithography

Optical lithography system consists of illumination, mask, projection lens system, and a wafer as shown in Fig. 1.1. Light beam delivered from a laser forms a certain illumination shape by diffraction optical element (DOE). A mask is illuminated by light through illumination lens, which makes intensity distribution on the mask uniform. The mask contains nanoscale mask images for patterns to be fabricated on a wafer. Diffracted light from the mask propagates through projection lens, which focuses the light on a surface of photoresist. In case of water immersion lithography, the space between the projection lens and photoresist is filled with water, whose

S. Shim and Y. Shin, *Physical Design and Mask Synthesis for Directed Self-Assembly Lithography*, NanoScience and Technology,
https://doi.org/10.1007/978-3-319-76294-4_1

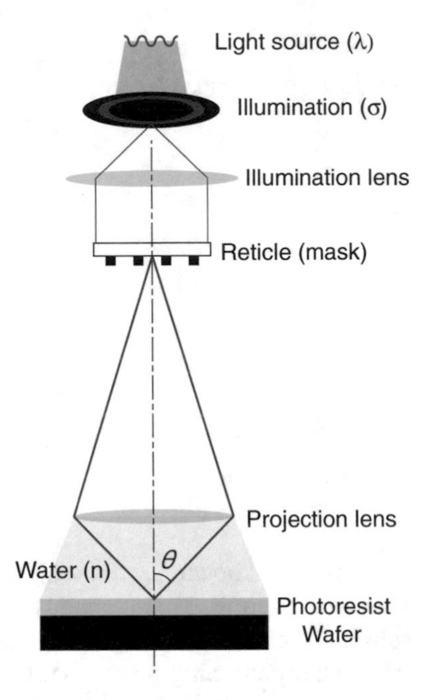

Fig. 1.1 Optical lithography system

refractive index is $n = 1.44$. In the region of photoresist that is exposed to the light, acid is generated by the interaction of light and photo-acid generator (PAG), and photoresist in that region is removed after develop process. Subsequent etch process removes substrate that is exposed through empty photoresist and so leaves a trench in substrate.

A minimum pattern size that the lithography system can support, namely resolution [1], is given by

$$\text{Resolution (min pitch/2)} = k_1 \frac{\lambda}{NA}, \tag{1.1}$$

where λ is the wavelength of light; NA is numerical aperture, whose definition is $n \sin \theta$, where θ is the maximum incident angle of light to photoresist surface and n is refractive index of material that fills the space between the projection lens and photoresist; k_1 is a constant and is equal to $\frac{1}{2(1+\sigma)}$, where σ is a partial coherence factor determined by illumination shape (a number of examples shapes are shown in Fig. 1.2).

A number of resolution enhancement techniques have been studied in regard to the parameters of (1.1).

- **Wavelength** (λ): Most intuitively, a light source of shorter wavelength can be used for better resolution, e.g., KrF (248 nm), ArF (193 nm), and EUV (13.5 nm). But this is an expensive choice because light source affects the whole lithography system including photoresist and projection lenses.

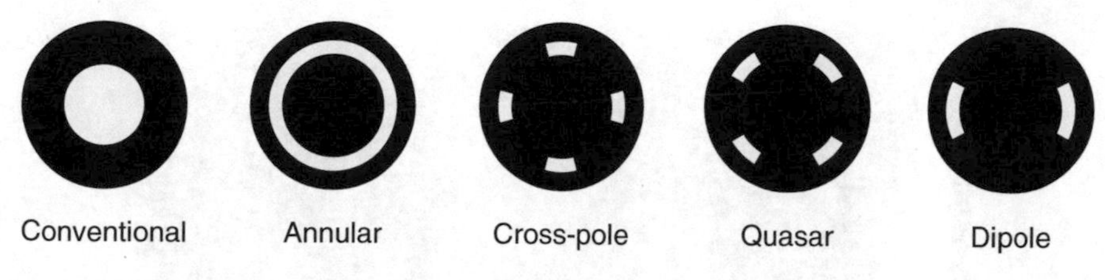

Fig. 1.2 Various illumination shapes

- k_1 **factor**: It is determined by the value of σ, which in turn is affected by the illumination shape. As a radius of bright region increases, σ also increases. For instance, σ becomes 1 (and k_1 becomes 0.25, which is in fact physical limit) if the radius of a ring is maximized in the annular illumination shown in Fig. 1.2, which is associated with the aggressive off-axis illumination (OAI). The value of σ can be controlled by adjusting axicon and zoom lenses [2].
- **NA**: Better resolution is possible with larger NA. NA is determined by optical elements, e.g., size of projection lens and physical apertures in the scanner. To take advantage of larger NA, new scanner system with higher performance optical elements should be developed. Larger NA (higher incident angle θ) however results in smaller depth of focus (DOF), which is a tolerance of placement error of wafer plane along the optical axis direction (i.e., normal to the image plane). DOF is defined by

$$DOF \propto \frac{1}{\sin^2 \theta}. \tag{1.2}$$

  Smaller DOF implies that the lithography system is more sensitive to the focus variation and is more likely to cause lithographic defect. Therefore, smaller DOF requires higher precision of process control. Immersion lithography is another way of increasing NA without sacrificing DOF: refractive index n increases (from 1 for air to 1.44 for water) instead of increasing θ; high-end immersion scanner system can support NA of 1.35 [2].

Figure 1.3 shows evolution of lithography technology for DRAM, NAND Flash, and Logic devices. Since standard optical lithography reaches its limit in about 20 nm technology node (see Logic in Fig. 1.3), aggressive RET techniques are adopted. However, their adoption is, most of the time, very difficult and delayed. For instance, k_1 factor has been pushed down to 0.25, which is its physical limit,[1] by adopting multiple patterning, which however is expensive. Liquid of high refractive index (e.g., 1.65 NA [3]) has been developed for higher NA, but it has not been adopted in production use yet due to high absorbance, high viscosity, and poor surface tension [4].

---

[1]Practical limit of k_1 is believed to be about 0.27 [2].

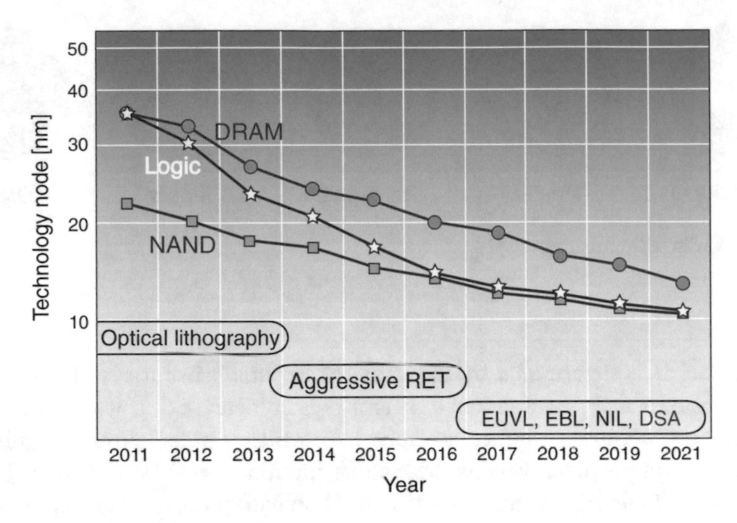

Fig. 1.3 Evolution of lithography technology [5]

1.2 Next Generation Lithography Technologies

To overcome the limitation of optical lithography, a few next generation lithography (NGL) technologies have been developed. Three examples of such technology are introduced in this section: extreme ultraviolet lithography (EUVL), electron beam lithography (EBL), and nanoimprint lithography (NIL).

1.2.1 Extreme Ultraviolet Lithography (EUVL)

EUVL uses radiation of wavelength 13.5 nm, which offers significant potential to extend the resolution limit of optical lithography. Most advanced 193 nm ArF immersion optical lithography can resolve features down to 38 nm with some resolution enhancement techniques such as off-axis illumination and water immersion projection lenses that have ultrahigh NA of up to 1.35. In contrast, 13.5 nm EUVL system has the potential to reduce the resolution to below 10 nm [6].

However, there are many challenges in developing EUVL system for high volume manufacturing (HVM) [7]. Because EUV radiation is absorbed by all materials including air, the optical elements responsible for imaging capability of EUV scanner have to use reflective lenses (mirrors) rather refractive lenses as shown in Fig. 1.4. It is difficult to manufacture high reflective mirror for EUV, therefore light power may be significantly degraded while light is reflected by many mirrors, which negatively affects throughput. For the same reason, the entire optical path from the light source to the wafer has to be in near-vacuum, which requires very expensive facility. Mask defects remain the most significant problem. Since reflective imaging system requires

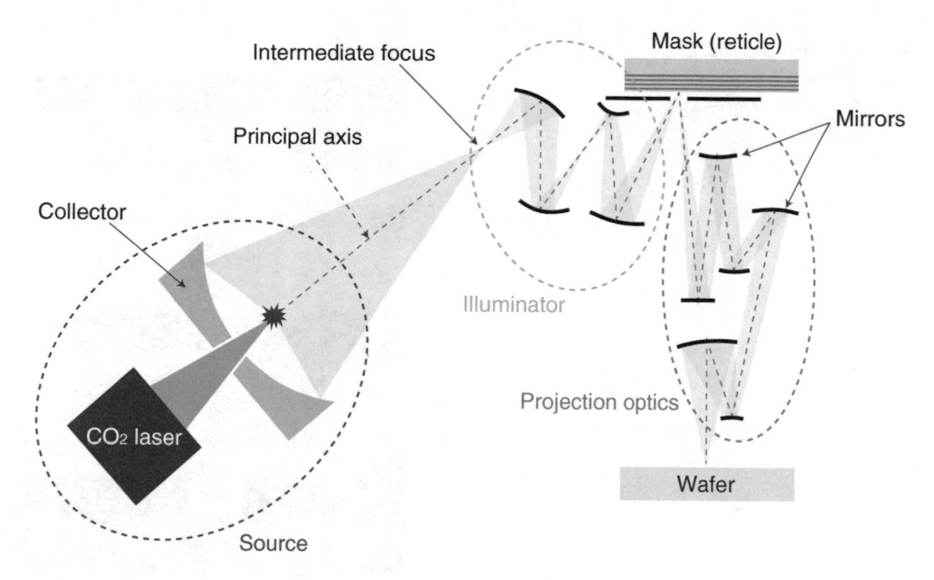

Fig. 1.4 EUVL system, which consists of the optics for collecting the light (collector), conditioning the beam (illuminator), and pattern transfer (projection optics) [8]

a reflective mask, which consists of multiple dielectric layers, it is very difficult to inspect and remove defects in such multiple layers.

1.2.2 Electron Beam Lithography (EBL)

EBL has been widely applied in construction of photomasks and templates for conventional HVM patterning technologies such as deep ultraviolet lithography (DUVL), EUVL, and NIL. Figure 1.5 shows a typical process of EBL: (a) substrate is coated with resist that is sensitive to electrons, (b) resist is exposed to scanning electron beam, and (c) resist in the exposed region is dissolved using a specific solvent (in case of positive lithography).

Recently, EBL itself has been proposed as a patterning solution of nanoscale structures on a wafer; it is called direct write electron beam lithography [9] in this application. Since a pattern is created by electrons, EBL is capable of very high resolution. EBL requires no mask due to scanning electrons and also enables to write a variety of pattern geometries, which is a significant advantage over other lithography techniques. It is also a flexible technique that can work with a variety of materials.

However, EBL has a few fundamental and practical limitations [11]. EBL tools are inherently slow because scanning e-beam is several orders of magnitude slower than optical lithography systems due to the serial pattern writing method; EBL with a single beam scanning takes about 8 h for one chip pattern, which is an unacceptable

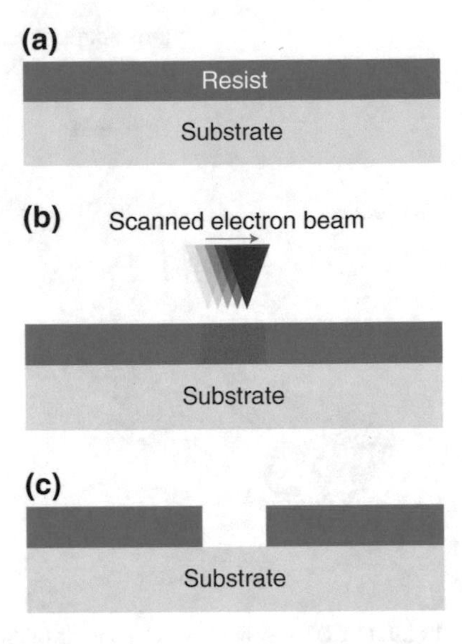

Fig. 1.5 EBL process: **a** resist coating, **b** exposure to scanning electron beam, and **c** development

throughput [12]. Some efforts have been put on EBL to improve its throughput. An example is variable shaped electron beam, but it still takes a few months to pattern on a 300 mm wafer with 65 nm minimum features [12]. Multiple e-beam technology [10] is another example, even though it has its own practical limitations as well. Cost is also a hurdle to EBL; EBL tools cost multimillion dollars and require frequent service to stay properly maintained. Finally, EBL sometimes causes damage on wafer substrate, which calls for more investigation and study.

1.2.3 Nanoimprint Lithography (NIL)

The general process of NIL is illustrated in Fig. 1.6. A template contains the nanoscale patterns to be fabricated on a wafer substrate (Fig. 1.6a). It is pressed into a polymeric material (resist) that has previously been deposited on the substrate (Fig. 1.6b). The template is filled with polymer, which is cured by UV light through the template. A copy of template shape is obtained in resist (Fig. 1.6c) and a residual layer of resist is then removed (Fig. 1.6d).

Because of very small proximity during NIL, mask data preparation (e.g., optical proximity correction and mask image fracturing) from design is not required, therefore NIL has the potential of cost-effective patterning compared to optical lithography and EUVL [14]. Because high resolution is expected for NIL template, high-quality electron beam writing is required for pattern generation. State-of-the-art nanoimprint lithography can be used for patterns down to 20 nm and below [15].

There are a few limitations of NIL. Lifetime of NIL template is not long due to its repeated physical contact with resist. So, template should be replicated to sustain

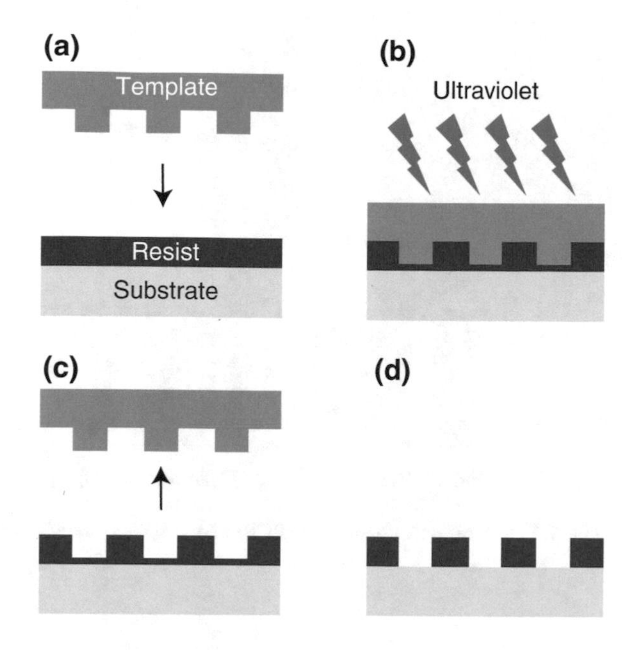

Fig. 1.6 NIL process: **a** a template with nanoscale patterns and resist on a wafer substrate, **b** a template is filled with polymer and cured by UV light, **c** template shape is obtained on a wafer, and **d** residual layer of resist is removed

the manufacturing process, and defect control is very important during replication. Due to its physical contact, NIL is susceptible to undesirable substrate distortion. Imprint process should be optimized to reduce such distortion. Imprint uniformity is important to keep the uniformity of critical dimension (CD) of patterns.

1.3 Directed Self-Assembly Lithography (DSAL)

Diblock copolymer (BCP) is a key material for DSA process. It is a unique string of two types of polymer as shown in Fig. 1.7a. One type of polymer is hydrophilic and another type is hydrophobic. Due to attractive and repulsive forces, they are arranged by themselves through microphase separation into various regular nanostructures such as cylinders, spheres, and lamellae [16] as shown in Fig. 1.7b. The feature size of these nanostructures is determined by the length of BCP and is not limited by the wavelength of light, which affects the feature size in optical lithography. In addition, pattering through self-assembly can be controlled by simple thermal annealing process, which significantly reduces the cost and improves the throughput. Out of all the variety of nanostructures shown in Fig. 1.7b, the cylindrical one is especially suitable for patterning contacts and vias in integrated circuits (ICs).

Since contacts and vias are not regularly located in practical layout as shown in Fig. 1.8a, a simple regular hole array generated by standard DSAL is not suitable for IC fabrication. Fortunately, topological guide pattern (GP) assisted DSA process has been proposed [17] to support patterning holes that are irregularly placed. Contacts

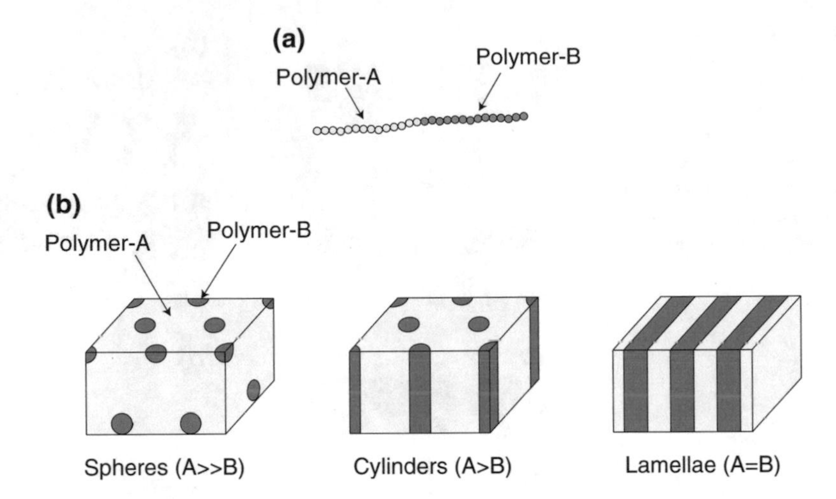

Fig. 1.7 **a** A diblock copolymer (BCP) and **b** typical morphologies determined by the relative composition of BCPs

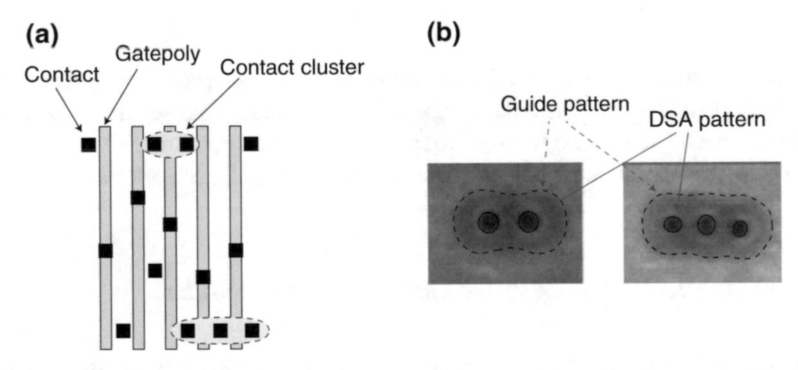

Fig. 1.8 **a** Contacts and contact clusters and **b** SEM images of corresponding GP and DSA pattern on a wafer

(or vias) that are close are grouped as shown in Fig. 1.8a and a unique surrounding contour, which serves as a GP, is identified for each group. GPs are patterned on a wafer through optical lithography, which implies that mask contains GPs not contacts. Each GP is filled with BCP; after heating and annealing, one type of polymer forms as contact holes, which are then etched away (see Fig. 1.8b). DSA process occurs within each GP region independently from other GPs.

There are two process integration methods for GP-assisted DSA. In Fig. 1.9a, a hole is patterned on a photoresist and is etched down to hardmask; it is then used as a GP. In Fig. 1.9b, a hole patterned on a photoresist is directly used as a GP. Patterning holes can be done through e-beam lithography or traditional optical lithography. It is well known that e-beam lithography exhibits high patterning fidelity but throughput is

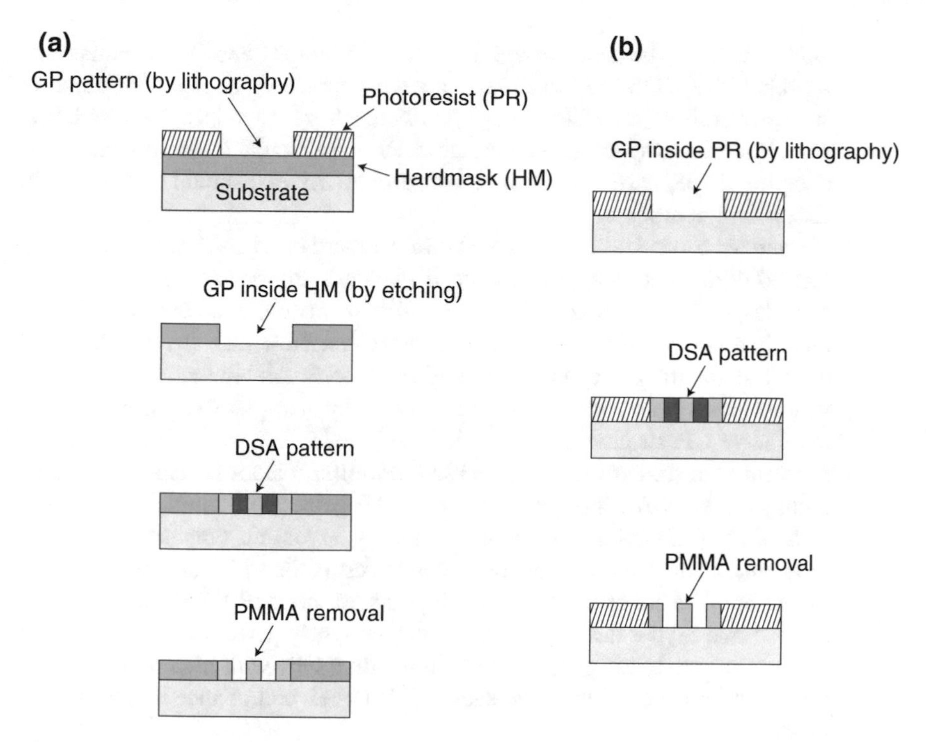

Fig. 1.9 Process integration for GP assisted DSA: **a** DSA inside photoresist and **b** DSA inside hardmask

very low. Integrated process of DSA and optical lithography, which is more popular, is often called DSA lithography (DSAL).

1.4 Overview of the Book

The remainder of this book consists of two parts. In part 1 that covers chapters from 2 to 6, physical design issues are studied. In particular, DSA defect probability is introduced in Chap. 2, which is a foundation for placement optimization and redundant via insertion in the following chapters. In part 2, the focus is mask optimization, specifically DSAL mask synthesis in Chap. 7, verification of guide patterns in Chap. 8, and cut optimization in Chap. 9.

1. **Physical Design Optimizations**

   - **DSAL manufacturabililty**: In DSAL, contacts (or vias) that are physically close are clustered together and patterned through GP. Small variations on GP contour, which may arise during lithography process, can lead to incorrect contact patterning in DSA process. A prediction of incorrect patterning, called

DSA defect, can be done through repeated lithography and DSA simulations. A probability of DSA defect is introduced in Chap. 2. A method of calculating such probability for all GP shapes is studied; a key is to reduce the calculation time. Experiments indicate that certain GPs, in particular those with large and complex shape, have high DSA defect probability and should be avoided in circuit design process.

- **Placement optimization for DSAL**: In standard cell based design, cells are located during automatic placement. If two cells are located side by side and some large GPs with high defect probability form across the two cells, placement is incorrect in manufacturing perspective. How that can be addressed during automatic placement is studied in Chap. 3. An alternative method, in which normal placement is assumed followed by post-placement optimization to fix large GPs, is also studied.
- **Placement optimization for MP-DSAL compliant layout**: In sub-7-nm technology node, DSAL is expected to be used together with multiple patterning technology (MP-DSAL). In such technology, layout decomposition is a key problem, in which contacts are clustered to form GPs which are then assigned to one out of a set of masks. Many layouts are not MP-DSAL compliant in a sense that layout decomposition is not perfectly performed leaving some GPs not correctly assigned to masks and some GPs with high defect probability. Optimization of layout to satisfy MP-DSAL compliance is presented in Chap. 4.
- **Redundant via insertion for DSAL**: In usual practice, more original vias are expected to be associated with redundant vias to increase via manufacturability. In DSAL, insertion of one redundant via may cause large number of vias to be clustered to form a GP. So, standard redundant via insertion problem should be reformulated such that large GPs with high defect probability do not form. This is studied in Chap. 5.
- **Redundant via insertion for MP-DSAL**: In MP-DSAL, redundant via insertion is even more complex because redundant vias are inserted while large GPs of high defect probability do not form and all GPs are correctly assigned to masks without any conflict. Chapter 6 covers this problem.

2. **Mask Synthesis and Optimizations**

- **DSAL mask synthesis**: Mask design for DSAL consists of two steps, inverse DSA and inverse lithography, which are addressed in Chap. 7. For a given contact or via layout, inverse DSA tries to find a set of ideal GPs that are expected to produce target contacts or vias. In inverse lithography, mask image (that contains GPs) is progressively refined so that ideal GPs can be patterned on a wafer. Inverse DSA and inverse lithography are extended to handle process variations.
- **Verification of GP**: GP shape is critical to successful patterning of contacts and vias in DSAL. Simulations may be used to check GP shape, but runtime is excessive. In Chap. 8, GP shape is characterized using a small number of geometric parameters. A binary verification function with those parameters

as inputs is constructed to give good or bad answer for a given GP shape. Specifically, a set of sample GPs is compiled. Each GP is represented by a vector of geometric parameters; it is submitted to DSA simulator to see whether its shape is good or bad; the result is attached to it as a tag. All GPs are located in a space of geometric parameters with their tags; the space is deformed so that a radial distribution is converted into the one in which the good and bad tags are separated by a hyperplane, which finally becomes a verification function.

- **Cut optimization**: Line-end cut process is used to create very fine metal wires in sub-14 nm technology. Cut patterns split dense line-space patterns into a number of wire segments, some of which become actual routing wires while the remainders are regarded as dummy. In sub-10 nm technology, cuts are smaller than optical resolution limit and DSAL or MP-DSAL is considered as an alternative patterning solution. In Chap. 9, cut optimization for DSAL is addressed. The goal is to relocate some cuts in such a way that cuts are grouped with each group being surrounded by a GP that is more favorable for manufacturing and wire extensions due to cut relocation are minimized; in MP-DSAL, each GP is assigned to one mask from a set of masks, and MP coloring conflicts should be minimized during assignment.

References

1. A. Wong, Resolution enhancement techniques in optical lithography. J. Quantum Electron. **47** (2001). (SPIE Press)
2. ASML, NXT:1950i User Manual (2011)
3. Y. Wang, T. Miyamatsu, T. Furukawa, K. Yamada, T. Tominaga, Y. Makita, H. Nakagawa, A. Nakamura, M. Shima, S. Kusumoto, T. Shimokawa, K. Hieda, High-refractive-index fluids for the next generation ArF immersion lithography, in *Proceedings of the SPIE Advanced Lithography* (2007), pp. 1–10
4. R.H. French, V. Liberman, H.V. Tran, J. Feldman, D.J. Adelman, R.C. Wheland, W. Qiu, S.J. McLain, O. Nagao, M. Kaku, M. Mocella, M.K. Yang, M.F. Lemon, L. Brubaker, A.L. Shoe, B. Fones, B.E. Fischel, K. Krohn, D. Hardy, C.Y. Chen, High-index immersion lithography with second-generation immersion fluids to enable numerical apertures of 1.55 for cost effective 32-nm half pitches, in *Proceedings of the SPIE Advanced Lithography* (2007), pp. 1–12
5. International Technology Roadmap for Semiconductors (ITRS), http://www.itrs2.net/
6. R.H. Stulen, D.W. Sweeney, Extreme ultraviolet lithography. J. Quantum Electron. **4**(5), 694–699, (1999)
7. B. Turkot, S.L. Carson, A, Lio, T. Liang, M. Phillips, B. McCool, E. Stenehjem, T. Crimmins, G. Zhang, S. Sivakumar, EUV progress toward HVM readiness, in *Proceedings of the SPIE Advanced Lithography* (2016), pp. 1–9
8. C. Wagner, N. Harned, EUV lithography: lithography gets extreme. Nat. Photon. **4**(1), 24–26 (2010)
9. H.C. Pfeiffer, Direct write electron beam lithography: a historical overview, in *Proceedings of the SPIE Advanced Lithography* (2010), pp. 1–6
10. B.J. Lin, Future of multiple-e-beam direct-write systems. J. Micro/Nanolithogr. MEMS MOEMS **11**(3), 1–6 (2012)
11. W.H. Cheng, J. Farnsworth, Fundamental limit of ebeam lithography, in *Proceedings of the SPIE Advanced Lithography* (2007), pp. 1–8

12. R.F. Pease, S.T. Chou, Lithography and other patterning techniques for future electronics. Proc. IEEE **96**(2), 248–270 (2008)
13. I. Yoneda, S. Mikami, T. Ota, T. Koshiba, M. Ito, T. Nakasugi, T. Higashiki, Study of nanoimprint lithography for applications toward 22nm node CMOS devices, in *Proceedings of the SPIE Advanced Lithography* (2008), pp. 1–8
14. K. Ichimura, K. Yoshida, S. Harada, T. Nagai, M. Kurihara, H. Hayashi, HVM readiness of nanoimprint lithography templates: defects, CD, and overlay, in *Proceedings of the SPIE Advanced Lithography* (2015), pp. 1–5
15. Y. Ootera, K. Sugawara, M. Kanamaru, R. Yamamoto, Y. Kawamonzen, N. Kihara, Y. Kamata, A. Kikitsu, Nanoimprint lithography of 20-nm-pitch dot array pattern using tone reversal process. Jpn. J. Appl. Phys. **52**(10R), 105201 (2013)
16. M. Muramatsu, M. Iwashita, T. Kitano, T. Toshima, M. Somervell, Y. Seino, D. Kawamura, M. Kanno, K. Kobayashi, T. Azuma, Nanopatterning of diblock copolymer directed self-assembly lithography with wet development. J. Micro/Nanolithogr. MEMS MOEMS **11**(3), 1–6 (2012)
17. H. Yi, Y. Bao, J. Zhang, C. Bencher, L. Chang, X. Chen, R. Tiberio, J. Conway, H. Dai, Y. Chen, S. Mitra, H.-S.P. Wong, Flexible control of block copolymer directed self-assembly using small, topographical templates: potential lithography solution for integrated circuit contact hole patterning. Adv. Mat. **14**(23), 3107–3114 (2012)

Part I
Physical Design Optimizations

Chapter 2
DSAL Manufacturability

In DSAL, contacts (or vias) are patterned in two steps. A GP that surrounds a group of nearby contacts forms on a wafer through optical lithography; member contacts then form through DSA process that arises within a GP. Final contact shape can be predicted by two sequential simulations: lithography simulation to predict GP shape and DSA simulation to predict contact. In reality, the two simulations should be repeated to account for lithography as well as DSA variations.

DSA defect probability is introduced, which is a probability that member contacts do not properly form and cause electrical short, open, etc. Computation of such probability involves many lithography and DSA simulations and so is a lengthy process. In standard cell based design, contact clusters (each cluster with its GP) that form within each cell or across two cells placed side by side can be identified; clusters with the same GP shape are grouped and representative defect probability for all clusters in the group can be computed beforehand. This allows defect probability to be accounted for while physical design such as placement and routing is performed.

2.1 DSA Defect

2.1.1 DSAL for IC Design and Fabrication

Application of DSAL to IC design and fabrication is illustrated in Figs. 2.1 and 2.2. Specifically, Fig. 2.1 corresponds to mask synthesis and Fig. 2.2 explains how final contacts are formed on a wafer.

For a given contact layout (or via layout), nearby contacts are grouped as shown in Fig. 2.1a. Grouping is based on BCP length and is performed in deterministic fashion [1]. For instance, when BCP length is 21 nm, two contacts whose

S. Shim and Y. Shin, *Physical Design and Mask Synthesis for Directed Self-Assembly Lithography*, NanoScience and Technology,
https://doi.org/10.1007/978-3-319-76294-4_2

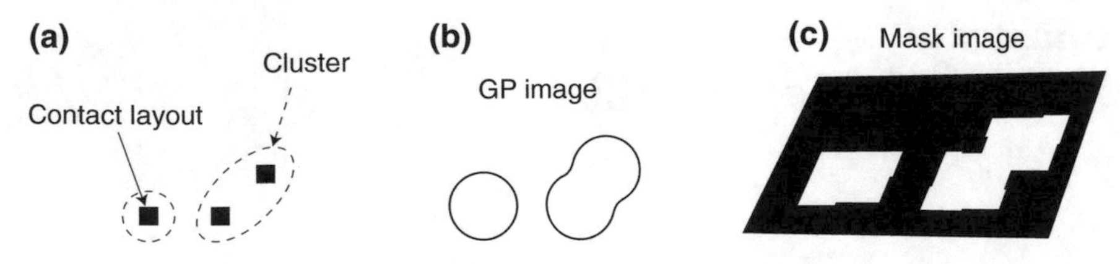

Fig. 2.1 Mask synthesis: **a** contact layout and clusters, **b** GP image, and **c** mask image after OPC

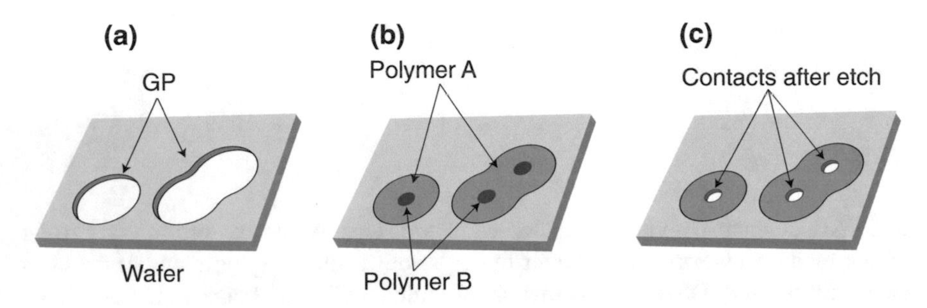

Fig. 2.2 DSA process: **a** GPs are patterned on a wafer through optical lithography, **b** GPs are filled with BCPs, and **c** contacts form after etch

center-to-center distance is between 35 and 50 nm have to be grouped together. For each group, the image of surrounding contour, namely GP image, is synthesized as shown in Fig. 2.1b. This process is named inverse DSA or GP synthesis, and is addressed in Chap. 7. The challenge lies in the fact that analytical synthesis method does not exist, so GP can only be obtained by repeated correction and checking. Finally, mask synthesis is performed to obtain a mask image of GPs (Fig. 2.1c). This is associated with standard optical proximity correction (OPC) to ensure that GPs patterned on a wafer using mask image in lithography process resemble target GP images in Fig. 2.1b as much as possible.

Actual GPs are created on a wafer (Fig. 2.2a) through optical lithography process such as 193 nm immersion lithography. GPs are usually sparsely located, so standard optical lithography is enough for their patterning. If a distance between some GPs is smaller than the pitch resolution of lithography, multiple patterning can be applied. Combination of DSAL and multiple patterning may be called MP-DSAL. A number of design issues in MP-DSAL are addressed in Chaps. 4, 6, and 9. Each GP is a large trench and is filled with BCP as shown in Fig. 2.2b. After heating and annealing, one type of polymer forms a shape of contact holes, which are then etched away (Fig. 2.2c). As a summary of DSAL, GPs are created through optical lithography and contacts are patterned through DSA process.

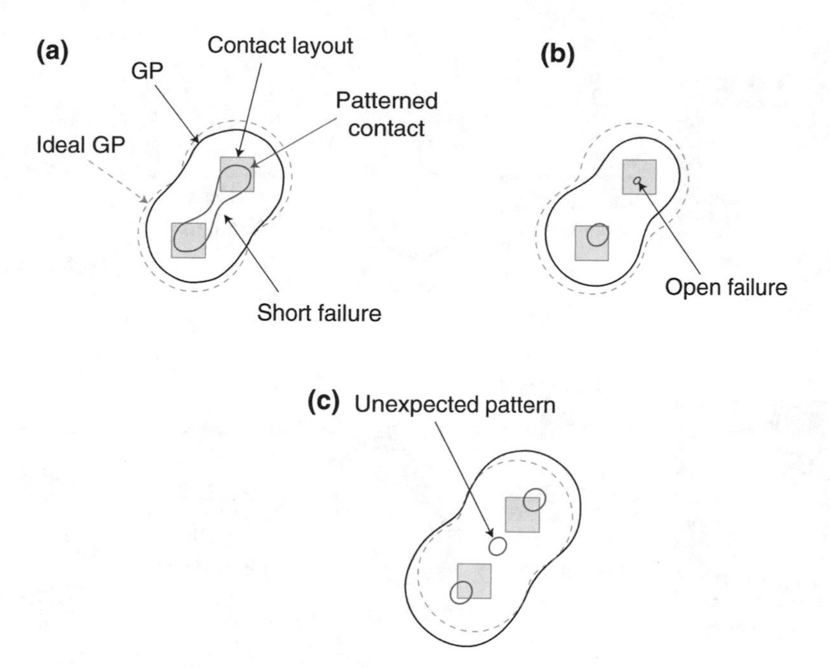

Fig. 2.3 Examples of DSA defect: **a** smaller GP may cause electrical short of contacts, **b** smaller and asymmetric GP may cause open failure, and **c** unexpected contact appears due to too large GP

2.1.2 Lithography-Induced DSA Defect

Even though an ideal GP and its mask image are synthesized, GPs patterned on a wafer may not be in ideal shape due to lithography variations caused by the errors in scanner focus, exposure energy, and mask manufacturing [2–4]. This in turn makes contacts to incorrectly form during DSA process and leads to DSA defect. Some defect examples are shown in Fig. 2.3. If a mask is exposed with lower illumination energy, GP may be patterned smaller than the target as shown in Fig. 2.3a; a smaller distance between adjacent contacts after DSA process may cause electrical short. A GP may be deformed asymmetrically due to different optical interferences with neighbor GPs as shown in Fig. 2.3b; one or more contacts may not open in this case. Scanner focus error may result in a stretch in GP as shown in Fig. 2.3c; the size of GP does not properly match BCP length and some unexpected contacts may appear. Since contacts are patterned through two independent processes, optical lithography and DSA, small variations on some critical GP contour may cause large error on final contacts.

DSA defect can be predicted through repeated lithography and DSA simulations. As illustrated in Fig. 2.4a, a lithography simulation with a mask image of GP yields an expected GP shape on a wafer, called a litho image; it is then submitted to DSA simulator [5, 6] and expected shape of contacts, called DSA image, is obtained. To account for lithography variations, the lithography simulation is repeated while

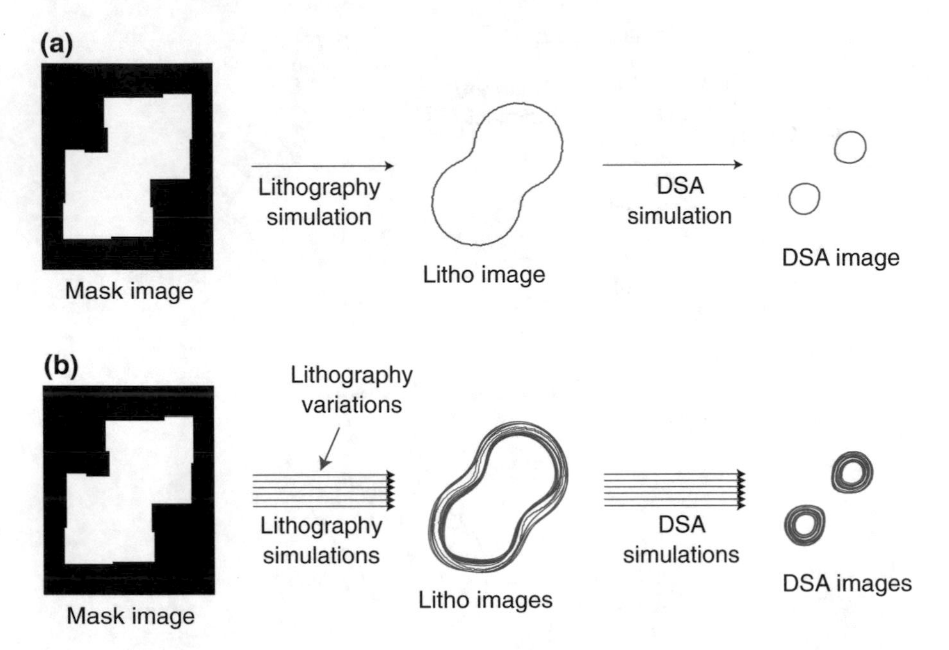

Fig. 2.4 **a** Mask image of a GP, expected shape of GP after optical lithography (litho image), and expected shape of contacts after DSA process (DSA image); **b** mask image of a GP, litho images under various lithography conditions, and corresponding DSA images

lithography parameters (e.g., scanner focus, exposure energy, and mask manufacturing error) are varied [2]; the result is a set of litho images as shown in Fig. 2.4b. Each litho image is then submitted to a DSA simulator, which outputs an DSA image. Each contact, in the end, is associated with multiple DSA images, and the size and location of the DSA images can be examined to verify whether DSA defect may occur.

2.2 DSA Defect Probability

2.2.1 Definition

DSA defect probability (or simply defect probability) quantitatively measures the sensitivity of a GP to lithography variations [8, 9]. The region bounded by the outermost and innermost contours of the multiple DSA images is called DSA process variation band (PVB), which is illustrated in Fig. 2.5. The size of DSA image is a main cause of the defect, and PVB indicates how much that size varies. If the size of DSA image is too small, contact is also expected to be too small on a wafer,

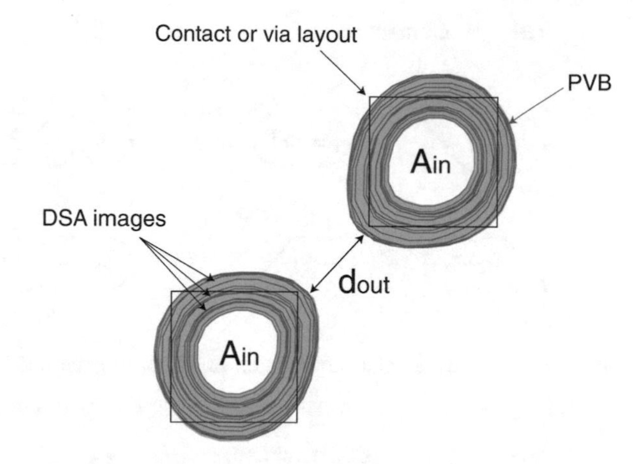

Fig. 2.5 DSA images and PVBs of two contacts

which may cause open failure. On the other hand, the distance of two DSA images determines whether a short failure may occur between two corresponding contacts.

Defect probability is now quantitatively defined as follows. Open failure is identified by examining the area of inner bound of PVB (see A_{in} in Fig. 2.5). Its probability is modeled by

$$D_{\text{open}} = \frac{M_A - A_{in}}{M_A - m_A} \times 100\%, \tag{2.1}$$

where M_A is an area of inner bound of PVB beyond which open failure never occurs (i.e., 0% defect probability), and m_A is an area of inner bound of PVB below which open failure occurs in 100%. The values of M_A and m_A are typically available from foundry fab [2, 3, 10, 11].

Defect probability of short failure is modeled by a distance between outer bounds of adjacent PVBs (see d_{out} in Fig. 2.5) and is given by

$$D_{\text{short}} = \frac{M_d - d_{out}}{M_d - m_d} \times 100\%, \tag{2.2}$$

where M_d and m_d are defined similarly to M_A and m_A, respectively. Defect probability of unexpected pattern shown in Fig. 2.3c can simply be defined. If unexpected pattern is observed in DSA images, the probability is 100%; otherwise, the probability is 0%. A GP corresponds to a group of contacts. The defect probability of GP (or a group of contacts) is defined by the maximum defect probability of its member contacts.

DSAL consists of two steps, optical lithography and DSA process. The latter is also affected by some variation sources. Unfortunately, there has been no study on those sources, and so DSA simulator with DSA process variations is currently unavailable. The PVB will become bigger if such simulator is available and is used, and defect probability (2.1) and (2.2) will then increase.

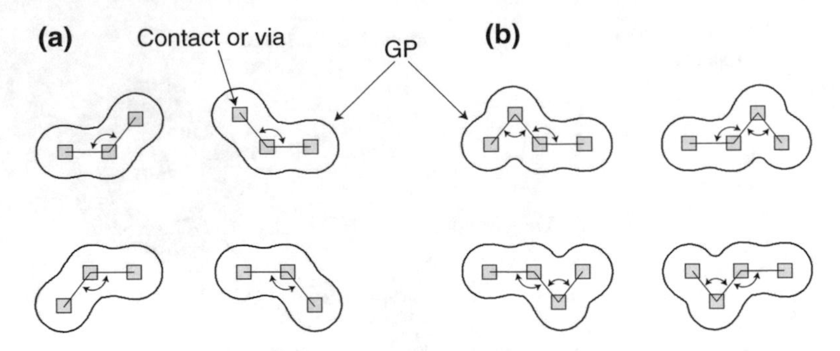

Fig. 2.6 An example of identical clusters of **a** three contacts and **b** four contacts

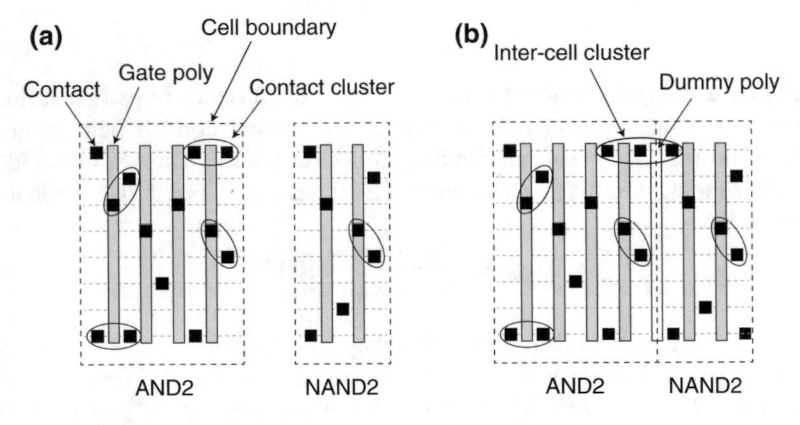

Fig. 2.7 **a** Contact clusters in individual standard cells and **b** intercell cluster after two cells are located side by side

2.2.2 Defect Probability Computation

Computation of defect probability involves multiple lithography and DSA simulations (see Fig. 2.4b), which are lengthy process. It is impractical to repeat the computation whenever a GP is encountered.

Fortunately, the variety of GPs that arise in the layouts is limited so the defect probability for each candidate GP can be calculated beforehand. This stems from the fact that GP is the same for the same number of contacts that are aligned in topologically same fashion as shown in Fig. 2.6. In addition, popular gridded design rule (GDR) [13] limits the position where contact or via can be located. For example, contact can be located directly on a gate poly or at the center of two adjacent gate polys when it is placed along the metal 1 track (see Fig. 2.7a).

In standard cell based design, contact clusters from all cells are listed while clusters with the same GP are grouped (see Fig. 2.7a). In addition, if two cells are located

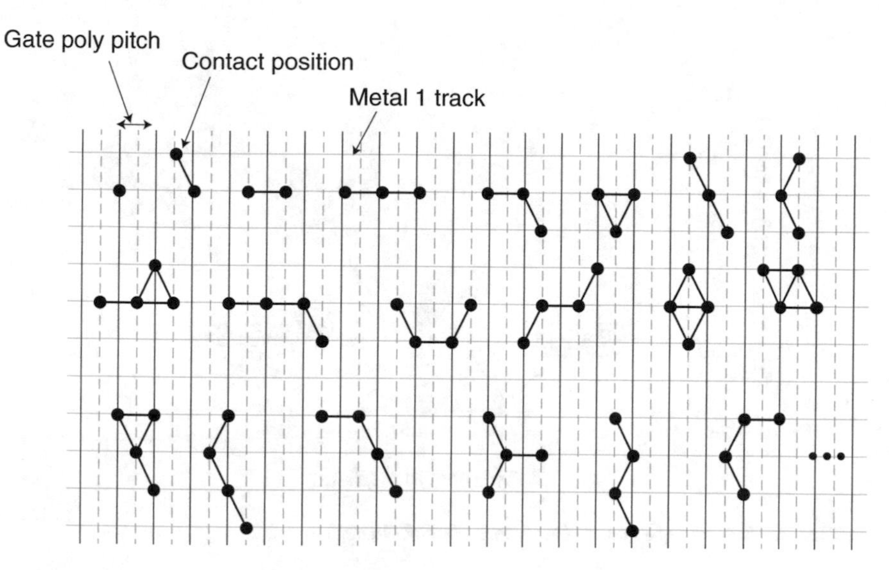

Fig. 2.8 Contact topologies of clusters that are extracted from standard cells

side by side as shown in Fig. 2.7b during placement, clusters that cross cell boundary, named intercell clusters, can form. All intercell clusters can also be identified and grouped beforehand. For example, if a library contains 100 standard cells, intercell clusters can form in 40000 (4×100^2) cell pair boundaries given that each cell can be flipped along its vertical axis. A few examples of contact clusters are listed in Fig. 2.8.

If a whitespace (of any width) exist between two cells, intercell clusters never form. Via is located where two metal tracks intersect. It has been observed [7] that unique via clusters are not many, e.g., only 31 clusters with four or less vias, while clusters of more than four have very high defect probability.

GP shape is affected by nearby GPs because GP is patterned on a wafer through optical lithography, which is affected by pattern density. For each unique cluster (of contact or via), its GP may be surrounded by dense or sparse neighbor GPs. For each surrounding, defect probability of a GP can be calculated; the one with the higher value may represent the probability for that GP (in conservative manner).

2.3 Experimental Observations

Defect probability is demonstrated using a 10-nm synthetic library. Sample layouts with 15-nm NanGate library [14] are shrunk so that they can follow the GDR of 10-nm technology, in which the minimum contacted poly pitch (CPP) is 45 nm, size of contact is 22 nm, and metal 1 track pitch is 36 nm. ArF immersion

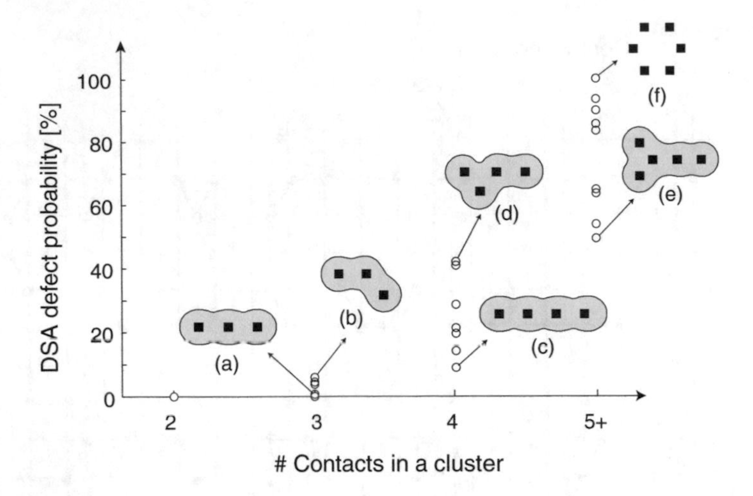

Fig. 2.9 DSA defect probability for various contact clusters

lithography (1.35 NA) with annular illumination is assumed, and self-consistent field theory (SCFT) based DSA simulator [5, 6] is used.

A simplified library contains 76 cells, so the total number of cell pair configurations is 23104 (4×76^2). Contacts that are within a distance of 57 nm are clustered, so two contacts with horizontal distance at CPP are always clustered. About 300 clusters are identified from individual standard cells, and 40000 intercell clusters are identified from all cell pairs; but unique clusters are only 41 (Fig. 2.8 illustrates 20 of those). Defect probability calculation takes about 70 h.

The experiments indicate that the defect probability increases as a cluster contains more contacts as shown in Fig. 2.9. There is a substantial variation of defect probability even for the clusters of the same number of contacts. Cluster (a), (c), and (e) have lowest defect probability in each category. This is because of symmetry in corresponding GP, which makes it less sensitive to lithography variations because BCP tends to be aligned periodically for lower energy. Due to lack of symmetry, cluster (b) and (d) have the highest defect probability. Cluster (f) is not associated with GP, since GP for such cluster does not exist, so its defect probability is 100%.

Defect probability computation in Sect. 2.2.2 is only approximate. This is because clusters with the same GP shape are grouped; actual computation is performed for only two clusters, one with dense and the other with sparse neighbor GPs; higher defect probability is considered to be the probability for all clusters in the group. In order to validate this approximate computation, 35 groups with each group consisting of five clusters are arranged. For each cluster, neighbor GPs are randomly located, and defect probability is calculated for its GP. Such probability is exact and identified on the y-axis of Fig. 2.10. For each group of five clusters, approximate probability is shown on the x-axis. It is indicated that the error of approximate defect probability decreases as exact defect probability is small. The approximate method predicts higher value than the exact method; this is safe in manufacturing perspective.

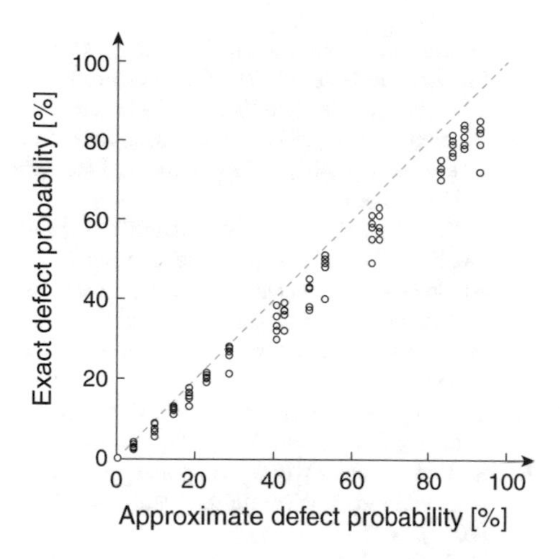

Fig. 2.10 Approximate versus exact defect probability

2.4 Summary

In DSAL, contacts and vias are formed through an annealing process, which occurs locally within each GP. Since GP is patterned via optical lithography process, its shape is affected by lithography variations. Small variations on some critical GP boundary may lead to incorrect DSA process and member contact holes may not properly form. DSA defect can be predicted through lithography simulations, while some variations are assumed, and subsequent DSA simulations. DSA defect probability can be defined as a function of distance between contact PVBs.

It has been shown that DSA defect probability can be computed for all possibility of GPs in a practical amount of time, since many GPs are identical each other. Experimental demonstrations have indicated that larger GP with complex shape has high DSA defect probability.

References

1. H. Yi, X. Bao, R. Tiberio, P. Wong, Design strategy of small topographical guiding templates for sub-15nm integrated circuits contact hole patterns using block copolymer directed self-assembly, in *Proceedings of the SPIE Advanced Lithography* (2013), pp. 1–9
2. L. Liebmann, S. Mansfield, G. Han, J. Culp, J. Hibbeler, R. Tsai, Reducing DfM to practice: the lithography manufacturability assessor, in *Proceedings SPIE Advanced Lithography* (2006), pp. 786–798
3. J.P. Cain, Design for manufacturability: a fabless perspective, in *Proceedings of the SPIE Advanced Lithography* (2013), pp. 1–9
4. K. Peter, R. Marz, S. Grondahl, W. Maurer, Litho-friendly design (LfD) methodologies applied to library cells, in *Proceedings of the SPIE Advanced Lithography* (2013), pp. 1–9
5. H.D. Ceniceros, G.H. Fredrickson, Numerical solution of polymer self-consistent field theory. Multiscale Model. Simul. **2**(3), 452–474 (2004)

6. N. Laachi, K.T. Delaney, B. Kim, S. Hur, R. Bristol, D. Shykind, C.J. Weinheimer, G.H. Fredrickson, Self-consistent field theory investigation of directed self-assembly in cylindrical confinement. J. Polym. Sci. Part B Polym. Phys. **53**(2), 142–153 (2015)
7. S. Shim, Y. Shin, Mask optimization for directed self-assembly lithography: inverse DSA and inverse lithography, in *Proceedings of the Asia South Pacific Design Automation Conference* (2016), pp. 83–88
8. S. Shim, W. Chung, Y. Shin, Defect probability of directed self-assembly lithography: fast identification and post-placement optimization, in *Proceedings of the International Conference on Computer Aided Design* (2015), pp. 404–409
9. W. Chung, S. Shim, Y. Shin, Redundant via insertion in directed self-assembly lithography, in *Proceedings of the Design, Automation and Test in Europe Conference and Exhibition* (2016), pp. 55–60
10. Samsung Electronics Corp. DFM Principal Engineer, *Personal Communication* (2015)
11. Samsung Electronics Corp. OPC Principal Engineer, *Personal Communication* (2015)
12. H. Yi, L. Azat, P. Wong, Computational simulation of block copolymer directed self-assembly in small topographical guiding templates, in *Proceedings of the SPIE Advanced Lithography* (2013), pp. 1–7
13. M.C. Smayling, V. Axelrad, 32nm and below logic patterning using optimized illumination and double patterning, in *Proceedings of the SPIE Advanced Lithography* (2009), pp. 1–10
14. Nangate 15nm open cell library, http://www.nangate.com/

Chapter 3
Placement Optimization for DSAL

In DSAL, each group of contacts (or vias) is associated with a GP. The number of member contacts and their relative position determine the shape of GP. Some GPs, for instance of those with asymmetric shape, have high DSA defect probability as addressed in Sect. 2.2, and have to be avoided. In standard cell based design, if such GPs are discovered within a cell, library designers are responsible to fix. If they are discovered at cell boundary when two cells are abutted after placement, a simplest approach to fix is to insert whitespace. It turns out that too many cell pairs require whitespace. In this chapter, two placement techniques are presented. In post-placement optimization [1], some cells are flipped and some two cells swap their position so that the number of cell pairs that require whitespace is minimized. In automatic placement, defect probability is taken care of while placement is performed. It is demonstrated that the two techniques yield layout of higher density (or less whitespace) by about 11–12% compared to layout of simple-minded approach.

3.1 Introduction

In DSAL, contacts that are closely located are patterned together in two steps:

1. **Lithography process**: A contour surrounding the contact cluster, called a guide pattern (GP), is synthesized on a mask [2, 3], which goes through optical lithography to create a GP image on a wafer.
2. **DSA process**: A GP is filled with block copolymers (BCPs), which are strings of two different types of polymer (one is hydrophilic and another is hydrophobic); BCP is arranged due to forces between the polymers and GP wall; one type of polymer is etched away, which leaves final contacts.

S. Shim and Y. Shin, *Physical Design and Mask Synthesis for Directed Self-Assembly Lithography*, NanoScience and Technology,
https://doi.org/10.1007/978-3-319-76294-4_3

Because contacts are created on a wafer through two independent steps, DSAL involves more defects than optical lithography does [1]. Some preliminary experiments show that about 1∼2% error in GP causes 4∼5% error in the final contact.

As discussed in Chap. 2, a cluster containing more contacts usually has higher defect probability because corresponding GP contour becomes more complex. To avoid such undesirable clusters, contact layout can carefully be modified in standard cell layout stages. Such clusters may be divided into smaller ones by relocating some contacts [4, 5]. However, when two cells are abutted (e.g., during placement), large clusters may still form across the cells as shown in Fig. 2.7. Verifying all intercell clusters[1] is very important but difficult. It may be one possibility to apply lithography and DSA simulations on whole layout, but impractical amount of time does not allow this approach.

Alternative would be to guarantee the correctness of intercell clusters in design time. Two key components for this approach are as follows:

- Determine defect probability of all cell pairs in advance, when two cells of each pair are located side by side.
- After standard placement, post-placement optimization is performed so that the use of whitespace can be minimized.[2]

For the first component, the intercell clusters of all cell pairs are identified. Fortunately, the number of unique clusters is quite manageable because there are many identical clusters. Each unique cluster is submitted to repeated lithography and DSA simulations; the simulation results are utilized to define the defect probability of a cluster, which is extended to define that of a cell pair. For post-placement optimizations, two methods are considered: (1) flip some cells (along y-axis) and (2) swap some adjacent cells as well as flip some. The second method benefits more but at the cost of increased wirelength due to larger cell displacement. Either method is applied to each placement row, and corresponding problem can be formulated as the shortest path problem of directed acyclic graph with nonnegative weight, which can be solved in linear time. The application of the methods on a few test circuits indicates 11% increase of placement density.

However, the post-placement optimization sometimes causes large cell displacement, which may increase total wirelength. Experiments indicate that total wirelength increases by 9%, which may degrade circuit timing. Instead of the post-placement optimization, defect probability may be taken care of while placement is being performed. A new defect probability-driven automatic placer is presented. It is implemented based on simulated annealing (SA) methods. Experimental demonstration indicates 12% higher placement density with only 3% increase of total wirelength.

[1]Intercell clusters do not form between the cells that are adjacent above and below due to the presence of power rails.

[2]Intercell cluster does not form if a whitespace of 1-poly pitch width is inserted.

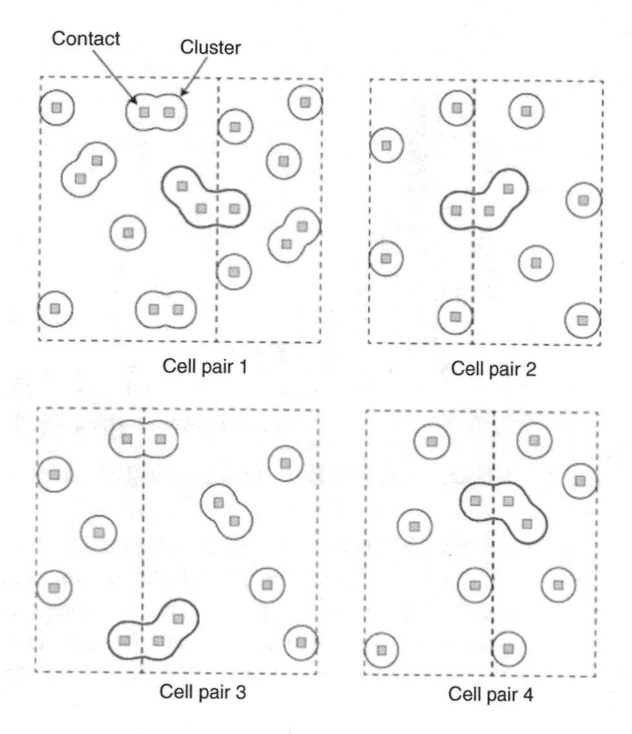

Fig. 3.1 Example cell pairs that contain identical intercell cluster

3.2 Defect Probability of Cell Pair

Since each standard cell is designed in a way not to cause a large and complex GP within the cell region, defect probability of a cell pair is determined by its intercell clusters. All possibility of cell pair combinations are tried, and their intercell clusters are extracted. The intercell clusters are grouped by examining similarity of topology of their member contacts. Each intercell cluster group is then mapped to cell pairs. For instance, four cell pairs shown in Fig. 3.1 have the identical intercell cluster (see red-colored clusters), thus are mapped to the same group. For each group, two cell pairs that have most dense (cell pair 1) and most sparse (cell pair 3) GPs are picked to take account most extreme interference of neighboring GPs and inputted to DSA defect probability computation. Defect probabilities of a cell pair are obtained as follows. GP and mask images of the cell pair are synthesized; lithography simulations with lithography variations [6–8] are performed for the whole GPs in the cell pair; subsequent DSA simulations [9, 10] are then performed only for a GP of the intercell cluster; defect probability of the intercell cluster is used as that of the cell pair. The larger value among the defect probabilities of those two cell pairs determines a representative defect probability of the group; in other words, all cell pairs in the group share the representative defect probability. Note that some cell pairs contain two (or more) intercell clusters (see cell pair 3 in Fig. 3.1); the final defect probability of such cell pair is determined by the largest value.

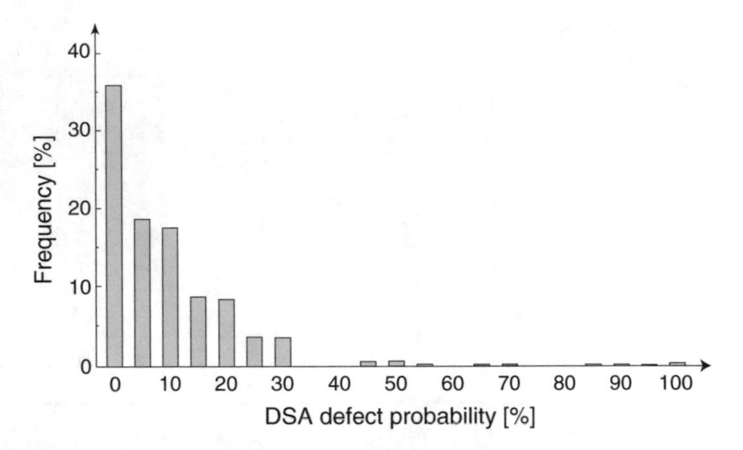

Fig. 3.2 A histogram of DSA defect probabilities of all cell pairs

Defect probabilities of all possible cell pairs are summarized as a histogram in Fig. 3.2. About 35% of pairs are defect free (i.e., zero defect probability): 17% pairs have no intercell clusters; 11% have 2-contact clusters, which have zero defect probability as shown in Fig. 2.9; the remaining 7% have clusters of three contacts being linearly aligned (cluster (a) in Fig. 2.9), which also have 0% defect.

3.3 Post-Placement Optimization

Suppose that a standard placement has been performed with some density target. A maximum defect probability that is allowed is given as a threshold (usually this will be 0%). If there is a cell pair being with high defect probability exceeds the threshold, a whitespace needs to be inserted between the cell pair so that no intercell cluster can form. The post-placement optimization aims to perturb the placement (as little as possible so that initial good placement is preserved) so that the amount of inserted whitespace is minimized. If whitespace needed for the post-placement optimization is not enough, initial placement may be performed again with lower density target (thus more whitespace) which is then followed by another post-placement optimization as shown in Fig. 3.3.

Two options are considered for post-placement optimization: cell flip, and cell flip together with swapping adjacent cells (see Fig. 3.4). Both options are applied to placement rows one by one. The second option will obviously yield better result, but at the cost of potentially increased wirelength due to larger cell displacement.

3.3.1 Cell Flipping

Let n cells in a row be denoted by C_1, C_2, ..., C_n as illustrated in Fig. 3.5a. The goal of cell flipping is to determine the orientation of each cell in such a way that

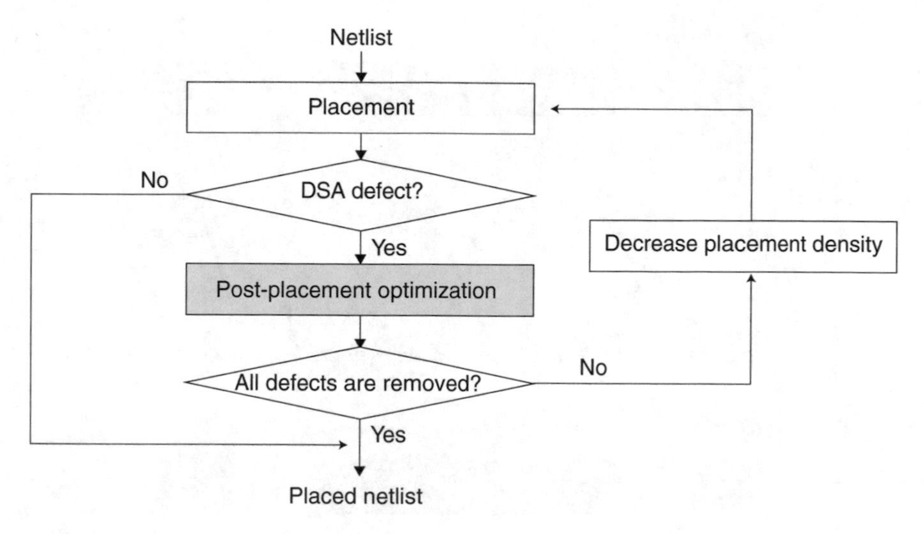

Fig. 3.3 Hypothetical placement flow

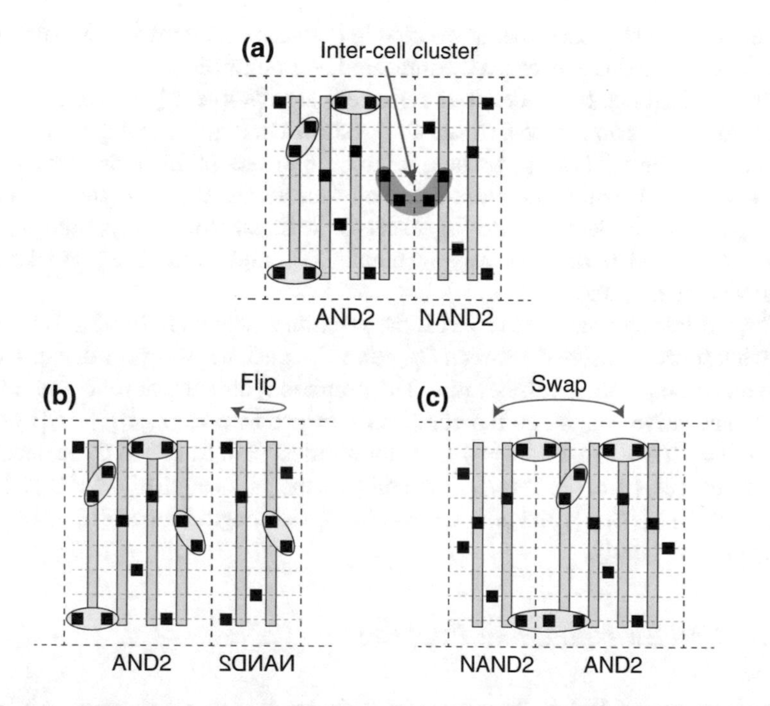

Fig. 3.4 **a** A cell pair contains undesirable intercell cluster, **b** one cell (NAND2) is flipped, and **c** positions of two cells are swapped

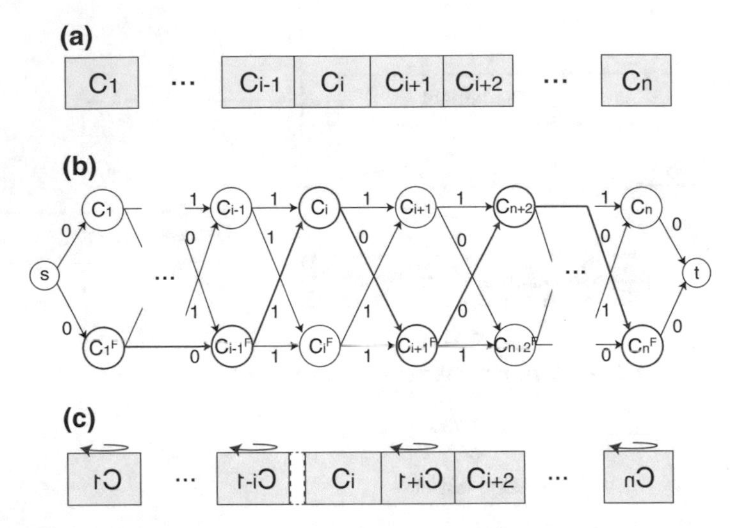

Fig. 3.5 **a** A placement row of n cells, **b** graph modeling to find optimal cell flipping, and **c** the result of flipping corresponding to the shortest path (marked red) of the graph

the number of whitespaces being inserted in between problematic cell pairs, whose DSA defect probability exceeds the threshold, is minimized.

The problem can be modeled as a shortest path problem [11]. Each cell corresponds to a set of two vertices, C_i and C_i^F, as shown in Fig. 3.5b, where superscript F indicates that the cell is flipped. There are four flipping combinations of two adjacent cells i and $i+1$, which are represented by four edges. If a configuration needs a whitespace, i.e., its defect probability exceeds the threshold, corresponding edge has a weight of 1; otherwise edge has 0 weight. The graph is finalized on adding two dummy vertices, s and t.

The shortest path from s to t yields the solution as shown in Fig. 3.5b and c; note that whitespace is inserted between C_{i-1}^F and C_i, since the weight value of the edge between corresponding vertices is 1. The graph is a directed acyclic graph (DAG) with nonnegative weight, so the complexity of problem is $O(|V|+|E|)$ or $O(n)$ since $|V|=2(n+1)$ and $|E|=4n$. Let the shortest path from s to C_i be denoted by a list of vertices, path(C_i). At C_{i+1}, we compare the path length of path$(C_i)\bigcup\{C_{i+1}\}$ and path$(C_i^F)\bigcup\{C_{i+1}\}$, and pick the path of shorter length for path(C_{i+1}); C_{i+1}^F can be obtained similarly.

3.3.2 *Cell Swapping and Flipping*

Let two adjacent cells be allowed to switch their positions in addition to flip themselves. The problem can also be cast to the shortest path of DAG. As shown in Fig. 3.6, either C_1 or C_2 or their flips can lead the placement row, so four vertices are fanouts of s vertex; similarly, four vertices are fanins of t vertex. Other position of placement row corresponds to six vertices since C_{i-1}, C_i, or C_{i+1} with their flips

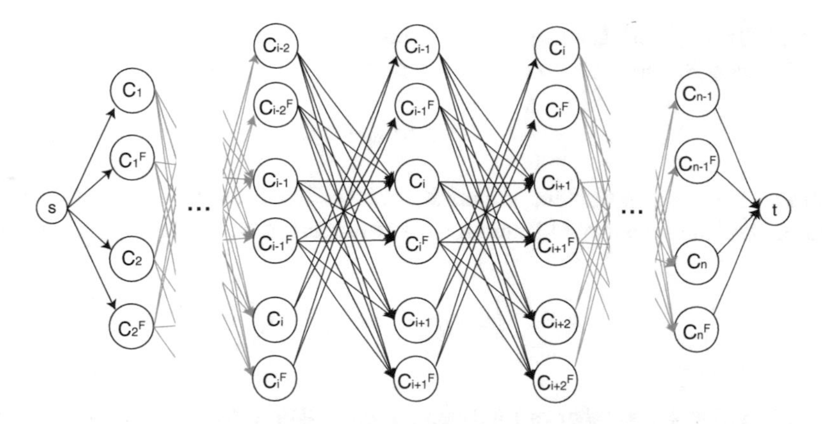

Fig. 3.6 Graph modeling to determine optimal configuration of cell swapping and flipping

can be located at ith position. The graph is then finalized by inserting relevant edges, e.g., C_i in $(i-1)$th position is connected to C_{i-1} and C_{i-1}^F in ith position (i.e., C_{i-1} and C_i are swapped) but it is not connected to C_i and C_i^F (and C_{i+1} and C_{i+1}^F) in ith position.

Cell swapping may be generalized so that cells can be relocated further in a row. However, the problem is not formulated as the shortest path of DAG anymore. Imagine C_1, C_2, C_3, C_4, and C_5 and assume that each cell can be relocated to one or two positions before and after its original position. If C_3 is relocated to first position, it may or may not have an edge to C_1 (or C_1^F): if C_2 moves to third or fourth position, C_1 can be located at second so the edge is allowed; if C_2 does not move, C_1 can only be located at third position so the edge will not be drawn. Furthermore, cell swapping incurs the increase of wirelength as demonstrated in Sect. 3.5 through experiments, so this generalized cell swapping will not be tried.

3.4 Automatic Placement

Instead of post-placement optimization combined with standard initial placement, we may take a unified approach in which DSA defect is taken care of while placement is being performed. This section addresses automatic placement problem, whose goal is to minimize both average defect probability of all adjacent cells and total wirelength. Maximum defect probability, which ultimately affects manufacturing yield, is also taken care of.

3.4.1 *Implementation of Placer*

A basic placer is based on simulated annealing (SA) [12]. A netlist is inputted into the new placer, which then generates an initial random placement (without any cell overlap). That initial placement is then gradually refined (while legality is maintained) during SA process. A cost of each placement is given by

$$\Gamma = \alpha C_d + \beta C_w + \gamma C_r, \tag{3.1}$$

where C_d is the average defect probability of all adjacent cells, and C_w corresponds to total wirelength estimated through half-parameter of bounding box. C_r is given by

$$C_r = \sum_{\forall \text{row } i} [WS(i) + RL(i) - RL_0]^2, \tag{3.2}$$

where $WS(i)$ is the number of total whitespace required between problematic cell pairs being with high defect probability over given threshold, $RL(i)$ is the sum of cell width, and RL_0 denotes the minimum $RL(i)$. Note that $WS(i) + RL(i)$ is the width that is required on row i and thus C_r in (3.2) guides all rows to receive necessary amount of whitespace.

- **Pre-placement**: Coefficients α, β, and γ are employed to treat three terms in (3.1) equally during cost evaluation. Their values are determined before SA loop starts and are given by

$$\alpha = \frac{\partial P}{\partial C_d}, \beta = \frac{\partial P}{\partial C_w}, \gamma = \frac{\partial P}{\partial C_r}, \tag{3.3}$$

where P denotes a placement instance. To obtain their values, the placement is perturbed (∂P) using operations that are employed in SA loop, and how much each term changes (i.e., the values of $\partial C_d/\partial P$, $\partial C_w/\partial P$, and $\partial C_r/\partial P$) is measured; this is repeatedly performed until the values converge. It is experimentally indicated that the coefficients converge very quickly as shown in Fig. 3.7, so its runtime is insignificant compared to the total runtime of placement.

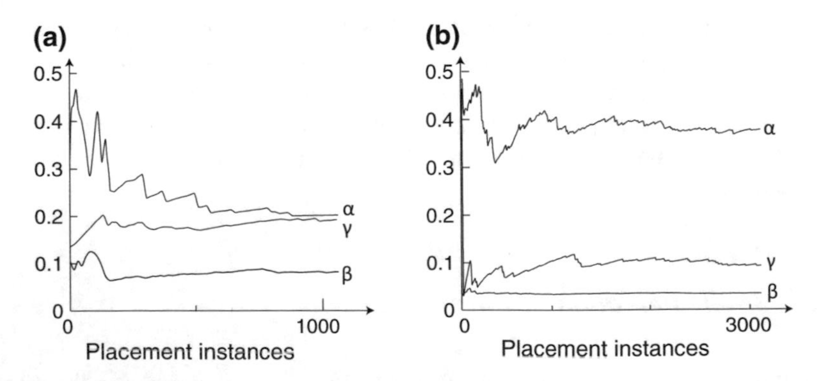

Fig. 3.7 Coefficient values of **a** tv80 and **b** b21

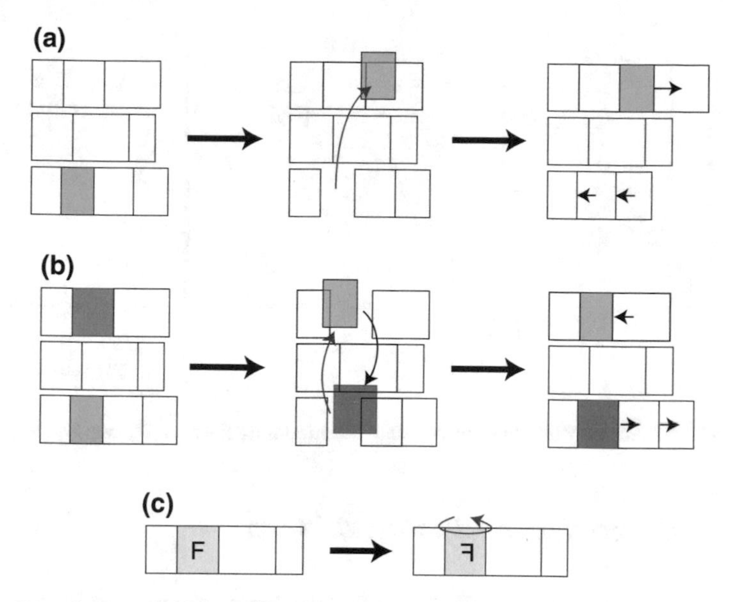

Fig. 3.8 Operations to generate a new placement instance: **a** Displace, **b** Swap, and **c** Flip

The number of cells in tv80 is about half number of cells in b21, so its C_d is about twice more sensitive to placement, which implies that α is half of that value in b21 as Fig. 3.7 indicates. The values of β and γ are inversely proportional to the width (or height) of placement region; tv80 occupies about half the area of b21, so its β and γ, respectively, are about $\sqrt{2}$ times larger.

- **SA-based placement**: Three operations are used to generate a new placement. "Displace" displaces a randomly picked cell to a randomly chosen new location (Fig. 3.8a); "Swap" randomly picks two cells and swaps their locations (Fig. 3.8b); "Flip" flips randomly picked cell along its y-axis (Fig. 3.8c). Cell overlap is removed by adjusting cell locations and cells are packed from left to right after each operation. At the beginning of SA loop, the three operations are chosen with equal probability, but as temperature decreases, "Displace" and "Swap" are chosen less frequently since they may affect C_w substantially.
- **Whitespace distribution**: During SA loop, cells are packed from left to right, so whitespace is placed only at the right end of each row. Once SA completes, in each cell row, adjacent cells being with high defect probability over given threshold are identified, and whitespace of single poly pitch width is inserted in-between. This is equivalent to assuming intercell margins between those cells, and polys can still be regularly placed for better lithography. If there are still some cell pairs being with high defect probability at the end of SA loop, placement fails and it should start all over again with lower placement density (or equivalently with more whiltespace) as is usually done when routing fails in standard placement and routing.

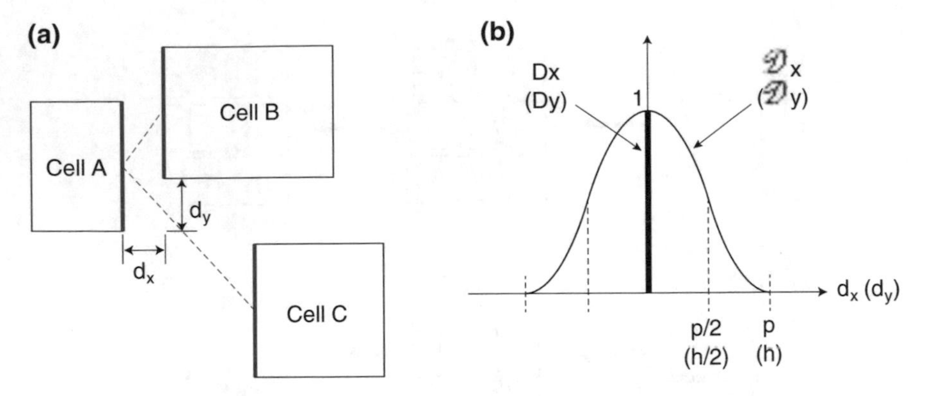

Fig. 3.9 **a** Three nearby cells and **b** the bell-shaped smoothing function $\mathcal{D}_x$ and $\mathcal{D}_y$

3.4.2 Considerations on Analytical Placer

Simulated annealing has been chosen for its simplicity of implementation. But, the similar (or same) idea can be applied to more sophisticated yet popular analytical placement method.

In the cost function (3.1), C_w can be modeled by differentiable function, e.g., logarithm sum exponential (LSE) approximation method [13–15]. C_d is discontinuous, but it can be made quadratic and differentiable. Imagine three cells A, B, and C as shown in Fig. 3.9a. The right boundary of cell A may touch the left boundary of B or C during a placement. For each cell pair, the exact defect probability is calculated by $D(i, j)D_x(d_x)D_y(d_y)$, where $D(i, j)$ is defect probability when the ith and jth boundaries are abutted, d_x and d_y are, respectively, the horizontal and vertical distances between the two boundaries, and D_x and D_y are direc delta functions, whose values are 1 when $d_x = 0$ and $d_y = 0$, and 0 otherwise (see Fig. 3.9b). Because D_x and D_y are not differentiable, they are approximated to bell-shaped functions [13, 15] $\mathcal{D}_x$ and $\mathcal{D}_y$. The form of the approximated function is given by

$$\mathcal{D}_x = \begin{cases} 1 - ad_x^2, & 0 \le |d_x| \le p/2 \\ b(d_x - p)^2, & p/2 \le |d_x| \le p \\ 0, & p \le |d_x|, \end{cases} \tag{3.4}$$

where $a = b = 2/p^2$ and p is one poly pitch width; d_y and h respectively are substituted for d_x and p in $\mathcal{D}_y$, where h is cell height. If two boundaries are physically close within a horizontal distance of one poly pitch and within a vertical distance of cell height (see cells A and B), they are likely to be abutted after legalization; otherwise, they are unlikely to touch each other, and defect probability is zero (see cells A and C). Therefore, the first term (C_d) of (3.1) can be expressed as sum of $D(i, j)\mathcal{D}_x(d_x)\mathcal{D}_y(d_y)$ for all i and j, which is called relaxed C_d and is still quadratic and differentiable. The last term C_r of (3.1) is not required in this analytical placer, because the first term forces to distribute whitespace indirectly. In summary, the cost function (3.1) can be integrated into the analytical objective function with slight modification.

The relaxed C_d is the same as the exact C_d when two cells are placed side by side or they are located apart enough to be free from pattern failure (i.e., zero defect probability). The two C_ds may be quite different during a global placement before legalization due to many overlapped and very close cells, whose exact C_d is 0 but relaxed C_d is larger than 0. To evaluate the error of the relaxed C_d, the exact and relaxed C_ds are compared for four test circuits (tv80, usb_func, aes_cipher, and vga_lcd). For a placement without legalization, it was shown that the relaxed C_d is 8.1% while the exact C_d is only 0.3% on average. But, both take the same value of 6.7% after legalization, which is only 1.4% smaller than the relaxed C_d for unlegalized placement. This implies that the relaxed C_d well represents C_d of legalized placement and so is good to use as a cost function during a global placement.

3.5 Experiments

A few test circuits from Open Cores [16] and ITC99 benchmarks [17] are used for experiments. The number of cells after logic synthesis ranges from 2k (b14) to 45k (ethernet). The post-placement optimization methods are implemented in Tcl scripts, which run on commercial physical synthesis tool [18].

Assessment of post-placement optimization: The two methods of Sect. 3.3 are assessed in terms of final placement density. A reference placement is constructed for comparison. After the standard placement of each circuit, placement rows are checked; whitespace is inserted between cell pair being with high defect probability over the threshold; if some rows have less whitespaces than necessary, placement is repeated with smaller density.

Placement densities are compared in Fig. 3.10a when the threshold is 0%. Cell flipping alone achieves 6% increase of density (from 82 to 88%) on average. Wirelength increases but very marginally by 2%, so this method perturbs an initial placement very little. Cell swapping and flipping allows 11% increase of density, which however comes at the cost of 9% increase of wirelength. Note that the wirelength is compared after actual routing.

Comparison with automatic placement: Post-placement optimization and SA-based placement are compared in placement density and total wirelength. SA-based automatic placement is repeatedly performed with increasing placement density while all defect probabilities are kept at 0%. Figure 3.11 shows that SA-based placement is slightly superior than post-placement optimization. It is also superior in total wirelength, which increases by 3% as opposed to 9% increase in post-placement optimization.

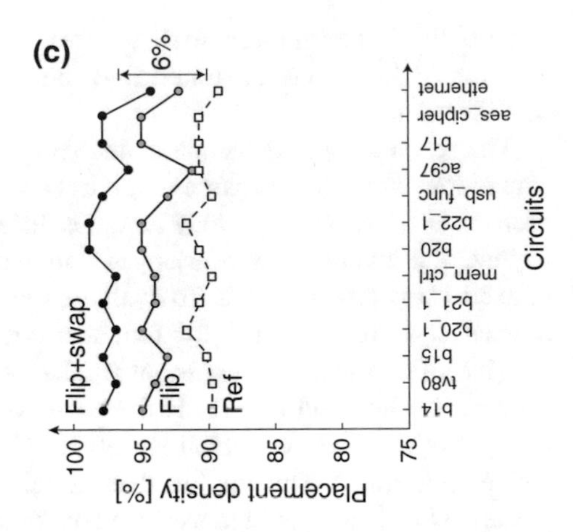

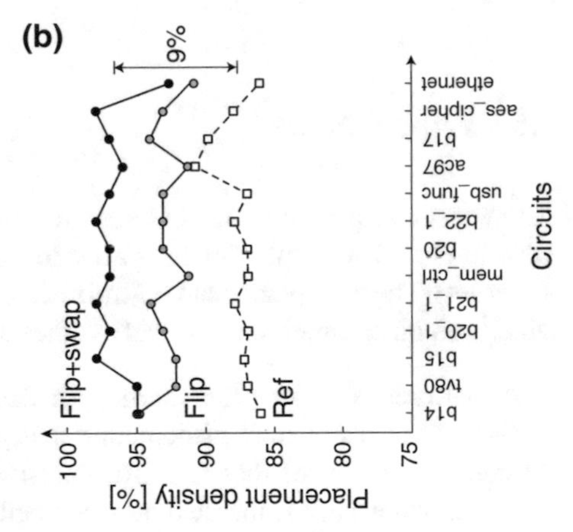

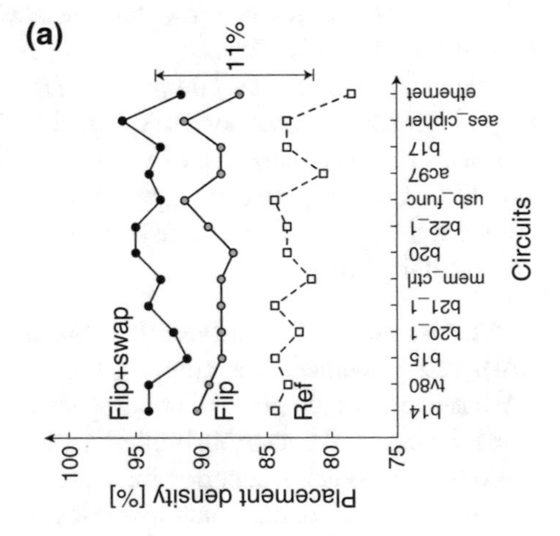

Fig. 3.10 Placement densities before (ref) and after (flip and flip+swap) post-placement optimization methods are applied, when the threshold of defect probability is **a** 0%, **b** 5%, and **c** 10%

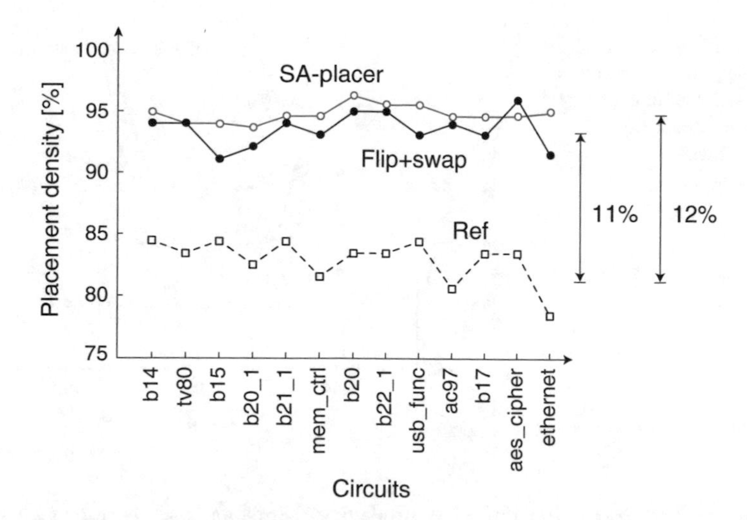

Fig. 3.11 Placement density of three methods: reference placement, post-placement optimization (flip and swap), and SA-based placement

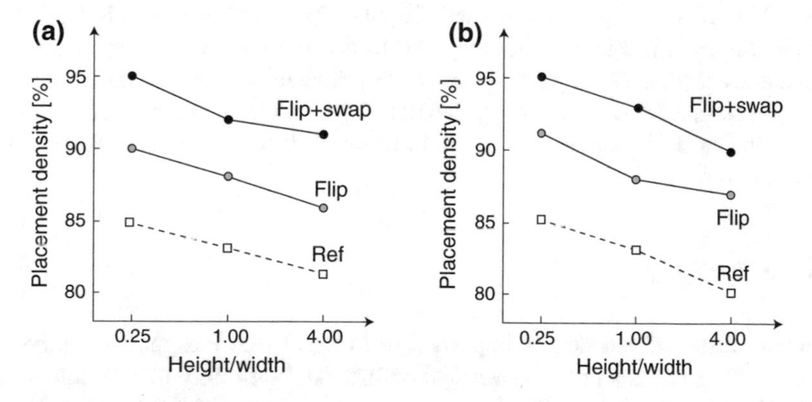

Fig. 3.12 Placement density with varying aspect ratio of placement region: **a** b20_1 and **b** mem_ctrl

Aspect ratio of a chip: Experiments in Fig. 3.10 have been performed assuming square placement region. Similar experiments of Fig. 3.10a are performed for two example circuits, with varying aspect ratio of region. Figure 3.12 shows the results. In the reference placement, the density decreases as the region becomes taller. This is because the width of region is determined by the widest row populated by cells and whitespaces (inserted between cell pairs of nonzero defect probability) and there are many rows (because the region is taller) having more whitespaces than necessary.

Initial placement is less perturbed by post-placement optimization as the region becomes taller, simply because each row contains less number of cells and so the optimization candidates also become sparse.

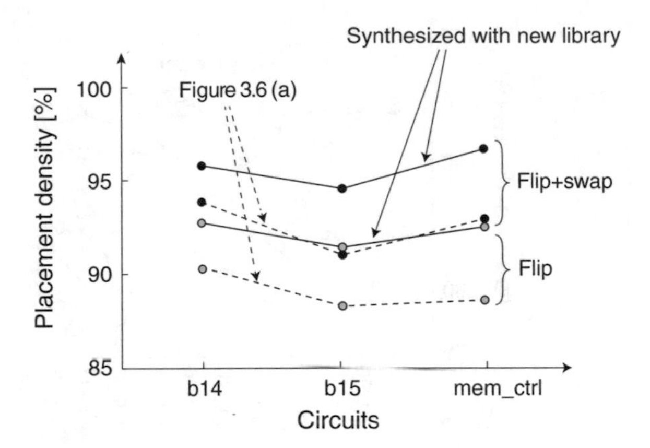

Fig. 3.13 Placement density (solid lines) when a circuit is synthesized with new library and is processed by post-placement optimizations

Library pruning using defect probability: More aggressive approach to take DSA defect into account would be to consider it early on from logic synthesis stage. For each library cell i, all cell pairs that include i in sample design are examined; if 70% or more cell pairs have nonzero defect probability, i is dropped from library. About 25% of cells (e.g., 1X_OR3 and 2X_AOI22) are removed in this process. All test circuits used in Fig. 3.10 are resynthesized using this new library; netlists are placed and are processed by post-placement optimizations. The resulting placement density is compared to that from Fig. 3.10a; comparison is illustrated for three sample circuits in Fig. 3.13. An average of 4% increase of density is observed (both in flip and flip+swap).

3.6 Summary

Intercell cluster of contacts is important in DSAL-based design. This is because corresponding GP has not been verified before. All possibility of intercell clusters can be identified in advance; they have been used to define the DSA defect probability of a cell pair in a library. A post-placement optimization has been addressed to minimize the amount of whitespace while all cell pairs are kept free from DSA defect; placement density increases by about 11%, but total wirelength increases 9% due to large cell displacement. Automatic placement, in which defect probability is considered together with total wirelength, has also been presented; placement density increases by about 12% with only 3% increase of total wirelength.

References

1. S. Shim, W. Chung, Y. Shin, Defect probability of directed self-assembly lithography: fast identification and post-placement optimization, in *Proceedings of the International Conference on Computer Aided Design* (2015), pp. 404–409

2. W. Wang, L. Azat, Y. Zou, T. Coskun, A full-chip DSA correction framework, in *Proceedings of the SPIE Advanced Lithography* (2014), pp. 1–11
3. L. Azat, G. Garner, M. Preil, G. Schmid, W. Wang, J. Xu, Y. Zou, Computational simulations and parametric studies for directed self-assembly process development and solution of the inverse directed self-assembly problem. Jpn. J. Appl. Phys. **53**(6) 06JC01–8 (2014)
4. H. Yi, X. Bao, R. Tiberio, P. Wong, Design strategy of small topographical guiding templates for sub-15nm integrated circuits contact hole patterns using block copolymer directed self-assembly, in *Proceedings of the SPIE Advanced Lithography* (2013), pp. 1–9
5. Y. Du, D. Guo, M. Wong, H. Yi, H. Wong, H. Zhang, Q. Ma, Block copolymer directed self-assembly (DSA) aware contact layer optimization for 10 nm 1D standard cell library, in *Proceedings of the International Conference on Computer Aided Design* (2013), pp. 186–193
6. L. Liebmann, S. Mansfield, G. Han, J. Culp, J. Hibbeler, R. Tsai, Reducing DfM to practice: the lithography manufacturability assessor, in *Proceedings of the SPIE Advanced Lithography* (2006), pp. 786–798
7. J.P. Cain, Design for manufacturability: a fabless perspective, in *Proceedings of the SPIE Advanced Lithography* (2013), pp. 1–9
8. K. Peter, R. Marz, S. Grondahl, W. Maurer, Litho-friendly design (LfD) methodologies applied to library cells, in *Proceedings of the SPIE Advanced Lithography* (2013), pp. 1–9
9. H.D. Ceniceros, G.H. Fredrickson, Numerical solution of polymer self-consistent field theory. Multiscale Model. Simul. **2**(3), 452–474 (2004)
10. N. Laachi, K.T. Delaney, B. Kim, S. Hur, R. Bristol, D. Shykind, C.J. Weinheimer, G.H. Fredrickson, Self-consistent field theory investigation of directed self-assembly in cylindrical confinement. J. Polym. Sci. Part B Polym. Phys. **53**(2), 142–153 (2015)
11. Y. Du, M. Wong, Optimization of standard cell based detailed placement for 16 nm FinFET process, in *Proceedings of the Design, Automation and Test in Europe Conference and Exhibition* (2014), pp. 1–6
12. C. Sechen, A. Sangiovanni-Vincentelli, The TimberWolf placement and routing package. IEEE J. Solid-State Circuits **20**(2), 510–522 (1985)
13. A. Kahng, Q. Wang, Implementation and extensibility of an analytic placer. IEEE Trans. Comput.-Aided Des. Integr. Circuits Syst. **24**(5), 734–747 (2005)
14. T. Chan, J. Cong, J. Shinnerl, K. Sze, M. Xie, mPL6: enhanced multilevel mixed-size placement, in *Proceedings of the International Symposium on Physical Design* (2006), pp. 212–214
15. T. Chen, Z. Jiang, T. Hsu, H. Chen, Y. Chang, NTUplace3: an analytical placer for large-scale mixed-size designs with preplaced blocks and density constraints. IEEE Trans. Comput. Aided Des. Integr. Circuits Syst. **27**(7), 1228 – 1240 (2008)
16. Opencores, http://www.opencores.org/
17. ITC99, http://www.cerc.utexas.edu/itc99-benchmarks/
18. Synopsys, IC Compiler User Guide (2015)

Chapter 4
Post-Placement Optimization for MP-DSAL Compliant Layout

Sub 7-nm technology node requires small contacts whose size and pitch are beyond optical resolution limit. Such fine features can be created by directed self-assembly lithography with multiple patterning (MP-DSAL) . In MP-DSAL, layout decomposition is a key problem, in which contacts that are physically close are clustered to form a GP which is then assigned to one of masks. Many practical contact layouts are not MP-DSAL compliant in a sense that layout decomposition is not perfectly performed leaving a few MP coloring conflicts and GPs of non-zero defect probability. This chapter introduces placement optimization to make a layout MP-DSAL compliant. The optimization problem is formulated as ILP, and a practical heuristic is also presented.

4.1 Introduction

Typical MP-DSAL process is shown in Fig. 4.1. Contacts that are physically close are grouped as a cluster (Fig. 4.1a). A contour surrounding each contact cluster, a guide pattern (GP) image, is synthesized (Fig. 4.1b). If two GPs are too close to create by traditional lithography, their mask images can be synthesized on different masks, which then go through MP process to form the GPs on a wafer (Fig. 4.1c and d). Each GP is filled with block copolymers (BCPs), which are self-arranged due to forces between the polymers and GP wall (Fig. 4.1e). One type of polymer (polymer-B) is then etched down to the substrate, which leaves final contacts (Fig. 4.1f).

In the basic DSAL without MP, contact clustering is uniquely determined [1, 2]. A larger and complex GP due to many member contacts is more likely to cause a DSA defect and so is difficult to manufacture [3]. MP-DSAL can mitigate this in a way as to split such large contact cluster into smaller ones, which are then assigned to different masks. Corresponding problem of clustering and mask assignment is called MP-DSAL decomposition problem and shown to be NP-complete [4–6]. Gupta

S. Shim and Y. Shin, *Physical Design and Mask Synthesis for Directed Self-Assembly Lithography*, NanoScience and Technology,
https://doi.org/10.1007/978-3-319-76294-4_4

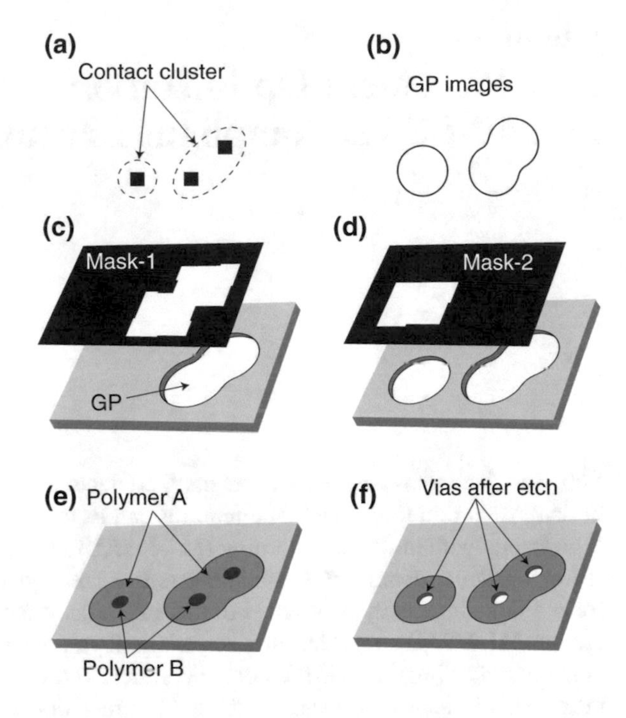

Fig. 4.1 MP-DSAL process: **a** contact clusters, **b** GP images, **c**, **d** MP process to create GPs on a wafer, **e** GPs filled with BCPs, which are then self-arranged, and **f** contacts created after polymer B is etched away

et al. address this problem based on ILP [5], which gives an optimum solution but is not scalable. Ivan et al. proposed a heuristic [7], which is to solve MP-DSAL decomposition for each cell row, but many color conflicts may still remain between adjacent cell rows.

To remove such color conflicts, it is required to modify contact layout by adjusting position of contacts that cause the conflicts. In this context, post-placement optimization problem is addressed. The goal of this problem is to perturb cell placement so that MP-DSAL decomposition can be successfully finished without any conflicts while the number of perturbed cells can be kept as small as possible. For a given library, MP-DSAL decomposition is performed for each cell, and all possible solutions are enumerated in advance. In each cell row of a given placement, some cells may be flipped to find the best placement that has no color conflicts between abutted cells in the row; this is modeled as finding paths in a directed acyclic graph (DAG). Final placement is then determined in a way that one placement is selected in every cell row while no color conflicts occur between adjacent cell rows and minimum number of cells are flipped; this problem is formulated as a graph-based fast heuristic, which is compared to ILP-based optimum solution.

4.2 MP-DSAL Decomposition

Contact grouping for DSAL is performed in deterministic manner, i.e., for a given contact layout, GPs are uniquely identified. In MP-DSAL, grouping is not deterministic anymore, and so is a difficult problem.

Let a pitch of two contacts be denoted by p. The minimum contact pitch supported by optical lithography is denoted by p_{litho}. If $p \geq p_{litho}$ (marked single mask in Fig. 4.2), two contacts are too far away to be clustered into a single GP; corresponding two GPs can be patterned using a single mask or they may belong to different masks.

Assume that p^+_{dsa} is the maximum contact pitch for two contacts to be clustered into a single GP. If $p^+_{dsa} \leq p \leq p_{litho}$, two contacts belong to different GPs, which have to be patterned using different masks. Let the minimum contact pitch supported by MP be denoted by p_{mp}; usually p_{mp} is smaller than p^+_{dsa}. If $p_{mp} \leq p \leq p^+_{dsa}$ (marked DSA/MP), two contacts may be clustered into a single GP or they may belong to different GPs, which reside on different masks. If $p^-_{dsa} \leq p \leq p_{mp}$, two contacts have to be clustered into a single GP. The values for p^-_{dsa} and p^+_{dsa} are determined by the natural length of BCP [1] that is employed; p_{mp} and p_{litho} are determined by lithography settings such as wavelength of light source, NA, and illumination type.

For a given contact layout, MP-DSAL decomposition problem is to group contacts and assign corresponding GPs to masks in such a way that all GPs are manufacturable with zero defect probability and mask assignment is performed with no (MP coloring) conflict. This problem has been shown to be NP-complete [5, 6]. ILP formulation [5] has been tried to solve the problem, which however can be applied only to a very small circuit.

In [6], this problem has been addressed in sequential fashion. In GP-MP approach, contacts are first clustered in the same manner as in standard DSAL, and mask assignment is then performed in a way that all clusters are assigned to masks so that the number of coloring conflicts is minimized. In MP-GP approach, on contrary,

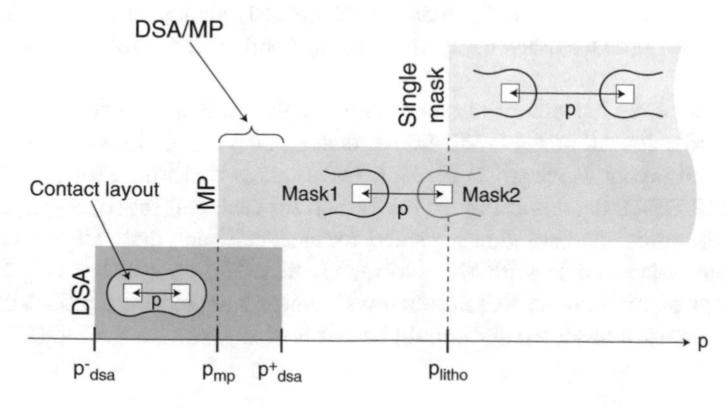

Fig. 4.2 Contact clustering and mask assignment for various contact pitch

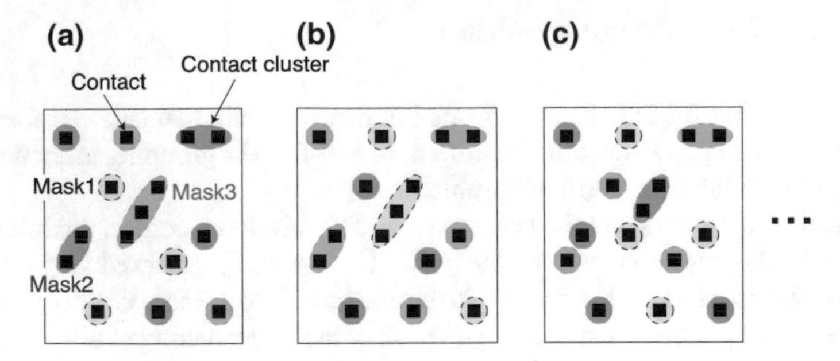

Fig. 4.3 Enumeration of MP-DSAL decomposition solutions for one standard cell: **a** and **b** for the same contact clusters with different colors, and **c** with different clusters

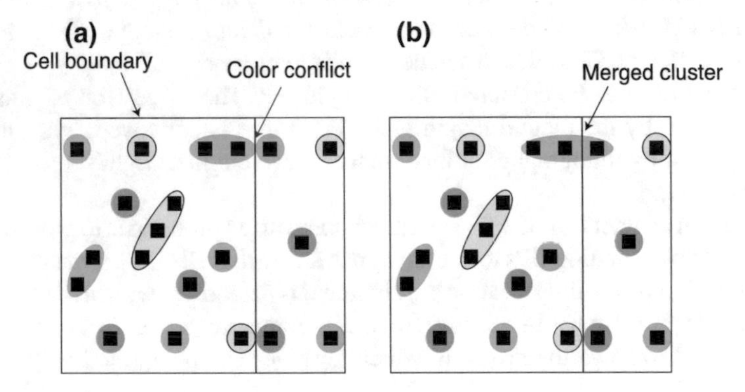

Fig. 4.4 **a** A color conflict between two close clusters and **b** removing the conflict by merging the clusters

mask assignment is first performed for contacts, and the contacts that are physically close and assigned to the same mask are then clustered together. In these approaches, contact clustering and mask assignment are performed independently, so the solution is far from the optimal leaving many MP coloring conflicts (this will be demonstrated in Sect. 4.4).

In [7], MP-DSAL decomposition problem is solved through solutions of individual standard cells. All possible MP-DSAL decomposition solutions are enumerated for each cell layout as shown in Fig. 4.3. Once this is performed for all cells in a library, MP-DSAL decomposition is performed for each cell row of a given circuit and one decomposition solution is picked for each cell such that cell row does not contain any adjacent cells with MP coloring conflict. This approach is very fast and gives an optimum solution for each cell row. However, MP coloring conflicts between cell rows are not considered and should be resolved as shown in Sect. 4.4).

4.3 Post-Placement Optimization

4.3.1 MP-DSAL Decomposition of Standard Cells

It is very difficult to solve MP-DSAL decomposition problem for whole contact layout because there are large number of possible contact grouping and mask assignment. In standard cell based design, MP-DSAL decomposition may be performed locally in each cell, which returns more than one decomposition solution. Decomposition of a whole design can be executed by choosing the solution of each cell and by combining them.

For each cell, all configurations of contact groping are first enumerated using only manufacturable cluster shapes being with 0% defect probability. For each configuration, all available mask assignments are also enumerated. A cell may have multiple configurations of contact grouping and mask assignment results, as shown in Fig. 4.3. This approach is exhaustive (it takes more than 10 h for 1,300 cells in a library), but is required only once for a library, and the result may be used for any circuits that are synthesized with the same library.

Two decomposed cells are then picked and abutted one by one, and it is checked whether there is a color conflict between the cell pair. This is performed for all possible cell pairs with all configurations of contact clustering and mask assignment, and the results are arranged as a table, which will be referred to in the post-placement optimization in Sects. 4.3.2 and 4.3.3. As a result, the size of the table is at most $4 \times (m \times n)^2$, where m is the number of cells in a library and n is the maximum number of clustering and mask assignment solutions of a cell. Note that, in some cell pairs (see Fig. 4.4), clusters causing a color conflict can be merged if they are close enough and the merged one is still manufacturable; such cell pairs are regarded as conflict free in the table.

4.3.2 Placement Optimization for Cell Row

For a given standard placement, each cell row is divided into small row segments in a way that all abutted cells are grouped and included in the same segment, as shown in Fig. 4.5a. As a result, there is one (or more) whitespace of one poly pitch width between every two adjacent segments, which do not cause color conflicts in-between. So, MP-DSAL decomposition problem can now be solved for each segment independently.

Let us assume a row segment consisting of five cells denoted by C_1, C_2, ..., C_5 as illustrated in Fig. 4.5a. The goal is to determine orientation of each cell in the segment so that MP-DSAL decomposition can be successfully finished. This problem is modeled using a DAG as shown in Fig. 4.5b. One cell position corresponds to two groups of vertices: possible decomposition solutions of the cell correspond to one vertex group (see white-colored vertices); and those when the cell is flipped correspond to another group (gray-colored vertices). An edge is then inserted between

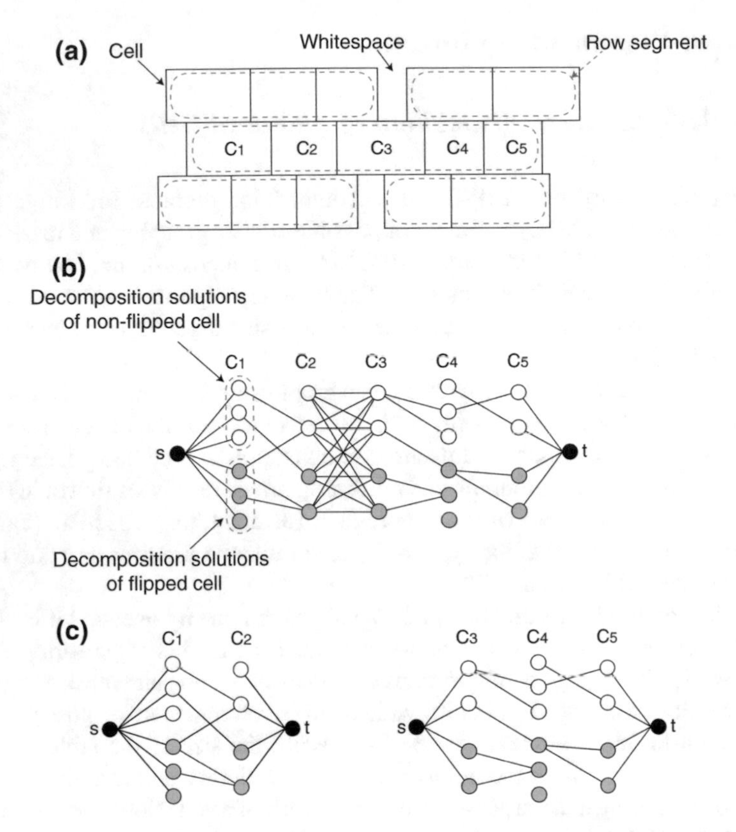

Fig. 4.5 **a** A standard placement, **b** a DAG of one cell row segment, **c** the DAG is split into small DAGs

two vertices from adjacent cell positions (e.g. C_1 and C_2) if the corresponding pair of decomposition solutions has no color conflict, which is identified by simply looking up the table constructed in Sect. 4.3.1; otherwise, no edge is inserted. Source and terminal vertices, s and t, are added finalizing the graph. All possible paths from s to t, which correspond to the best placements (i.e., best combinations of cell orientations and decomposition solutions of the member cells) of the row segment, are then found.

In some cases, vertices from two adjacent cell positions are fully connected (see C_2 and C_3 in Fig. 4.5b). This means that the cell pair is always conflict free regardless of what decomposition solution is chosen. This allows to split the graph into two small ones as shown in Fig. 4.5c; accordingly, corresponding row segment is also split into two small ones, which reduces complexity of the problem.

Table 4.1 Notations of ILP formulation

P_{ij}	jth best placement of ith row segment
F_{ij}	The number of flipped cells in P_{ij}
C_{ijkl}	A pair of P_{ij} and P_{kl} being with interrow color conflict ($i \neq k$)
x_{ij}	1 if P_{ij} is selected, 0 otherwise

4.3.3 Considerations of Interrow Conflict

There may still be some color conflicts between the best placements of adjacent segments in different cell row, so-called interrow conflicts. To finalize post-placement optimization, one best placement needs to be carefully chosen for each segment while no interrow conflicts occur and number of flipped cells is minimized. This problem is formulated as an ILP, which will be used as a reference for comparison to a graph-based heuristic for a circuit of practical size.

ILP formulation: Table 4.1 shows the notations that are used for ILP formulation. The problem is now to determine the value of x_{ij} with goal of minimizing the sum of F_{ij} for selected P_{ij}:

$$\text{Minimize} \sum F_{ij}x_{ij}, \quad \forall\, i \text{ and } j \tag{4.1}$$

subject to:

$$\sum_{\forall j} x_{ij} = 1, \qquad \forall\, i \tag{4.2}$$

$$x_{ij} + x_{kl} \leq 1. \quad \forall\, C_{ijkl} \tag{4.3}$$

Equation (4.2) constrains to select only one best placement in each row segment. Inequality (4.3) avoids that two best placements from different row segments have interrow conflict in-between.

Heuristic algorithm: A conflict graph is constructed as follows. A vertex corresponds to a best placement of each row segment, and the number of flipped cells in the placement determines its vertex weight, as shown in Fig. 4.6a. Vertices originated from the same segment belong to the same vertex group and have an edge in-between (see black edge). In addition, if two placements have interrow conflict, corresponding two vertices have an edge in-between (see red edge). These two edges correspond to (4.2) and inequality (4.3) in ILP formulation, respectively. The goal is now to pick one vertex in every vertex group while there are no edges between the picked vertices and sum of their weight values is minimized. The problem corresponds to finding all maximum independent sets (MISs) of the conflict graph, and the result corresponds to placements being with no color conflict at all. One with the smallest weight sum,

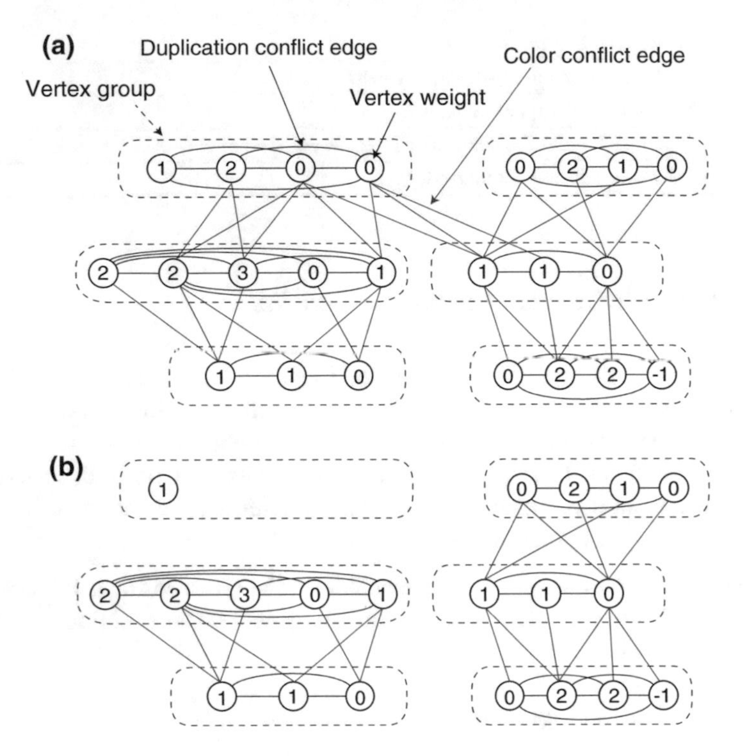

Fig. 4.6 Conflict graph **a** before and **b** after partitioning

corresponding to the conflict-free placement with minimum cell flipping, is selected as a solution.

The input graph is partitioned because finding MIS is an NP-hard problem and the graph is usually large. The partitioning is performed in a greedy fashion as follows. In each vertex group, a vertex being with no color conflict edge is first picked as a member of MIS; accordingly, other vertices in the group are removed together with their connected edges, as shown in Fig. 4.6b. Note that if a group has multiple vertices without color conflict edge, one with smallest weight value is picked, and if a group has no such vertex without color conflict edge, the group is skipped. This is repeated until no group has such vertex without color conflict edge. The algorithm is then applied to each partitioned graph, which results in MISs; they are then merged and correspond to an optimum placement.

4.4 Experiments

The ILP formulation and heuristic algorithm of Sect. 4.3.3 are implemented using Perl script with GUROBI [8] as an ILP solver. Some test circuits from Open Cores [12] and ITC99 benchmarks [13] are synthesized with 28-nm node industrial library. Contact

Table 4.2 Comparison of four MP-DSAL decomposition approaches. Triple patterning is assumed

Circuits	# Contacts	MP-DSAL decomposition without cell flipping		MP-DSAL decomposition with cell flipping			
		Row-by-row [7]	ILP [5]	ILP [15]		Heuristic [15]	
		# Conflicts	# Conflicts	# Conflicts	# Flipped cells	# Conflicts	# Flipped cells
spi	9169	17	0	0	0	0	0
mem_ctrl	10288	53	0	0	0	0	0
usb_func	10708	174	0	0	0	0	17
b14	14782	153	4	0	29	0	43
b21	39813	328	26	0	153	0	195
b22	72202	386	Time-out	0	208	0	312
b18	145452	502	Time-out	Time-out	–	0	528
ethernet	301423	891	Time-out	Time-out	–	0	782
b19	519046	1138	Time-out	Time-out	–	0	1026

layout is then appropriately shrunk and modified to follow virtual 7-nm design rule, which is assumed; contact size is 15 nm by 15 nm, and poly and metal track pitch are set to 44 and 35 nm, respectively. Test layout contains contacts from 10k (spi) to 500k (b19) as shown in Table 4.2. Triple patterning with DUV immersion lithography is assumed, and corresponding simulation settings are used for computing DSA defect probability [9–11] Referring to Fig. 4.2, p_{dsa}^{-} and p_{dsa}^{+} are respectively set to 35 and 55 nm, and p_{mp} and p_{litho} are assumed to be 50 and 120 nm, respectively.

Four MP-DSAL decomposition methods are compared in Table 4.2. Many MP coloring conflicts remain in row-by-row method (column 3), since interrow conflicts are not taken into account. ILP (column 4) completes the decomposition without or with very little coloring conflicts, but only in five small circuits due to its runtime limitation. ILP formulation in Sect. 4.3.3 returns a solution with no coloring conflicts as shown in column 5; the number of flipped cells is shown in the next column. The result of the heuristic algorithm of Sect. 4.3.3 is shown in the last two columns. Coloring conflicts are removed in all circuits even though the number of flipped cells is a little larger.

4.5 Summary

Sub 7-nm technology requires very small contacts, and MP-DSAL is one of the solutions to pattern such contacts. In MP-DSAL, decomposition is a key problem, which is to determine contact cluster and mask assignment. Existing MP-DSAL decomposition methods without layout optimization leave many color conflicts because layout is not MP-DSAL compliant. This chapter addressed post-placement optimization, whose goal is to flip minimum number of cells so that MP-DSAL decomposition can successfully be performed. All possible decomposition solutions are enumerated for each standard cell in advance, which are referred to during post-placement optimization for each cell row. One best cell flipping is determined in every cell row in a way that no color conflicts occur between abutted cells and adjacent cell rows while cells are minimally flipped. Corresponding ILP formulation and a graph-based heuristic algorithm have been presented and demonstrated in 7-nm technology.

References

1. H. Yi, X. Bao, R. Tiberio, P. Wong, Design strategy of small topographical guiding templates for sub-15nm integrated circuits contact hole patterns using block copolymer directed self-assembly, in *Proceedings of the SPIE Advanced Lithography* (2013), pp. 1–9
2. L. Azat, G. Garner, M. Preil, G. Schmid, W. Wang, J. Xu, Y. Zou, Computational simulations and parametric studies for directed self-assembly process development and solution of the inverse directed self-assembly problem. Jpn. J. Appl. Phys. **53**(6), 06JC01–8 (2014)
3. S. Shim, W. Chung, Y. Shin, Defect probability of directed self-assembly lithography: fast identification and post-placement optimization, in *Proceedings of the International Conference on Computer Aided Design* (2015), pp. 404–409
4. W. Wang, L. Azat, Y. Zou, T. Coskun, A full-chip DSA correction framework, in *Proceedings of the SPIE Advanced Lithography* (2014), pp. 1–11
5. W. Wang, L. Azat, Y. Zou, T. Coskun, A full-chip DSA correction framework, in *Proceedings of the SPIE Advanced Lithography* (2014), pp. 1–11
6. Y. Badr, A. Torres, Y. Ma, J. Mitra, P. Gupta, Incorporating DSA in multipatterning semiconductor manufacturing technologies, in *Proceedings SPIE Advanced Lithography* (2015), pp. 1–8
7. Z. Xiao, C. Lin, M. Wong, H. Zhang, Contact layer decomposition to enable DSA with multiple-patterning technique for standard cell based layout, in *Proceedings of the Asia South Pacific Design Automation Conference* (2016), pp. 95–102
8. Gurobi Optimization, Inc., Gurobi optimizer reference manual, http://www.gurobi.com/
9. L. Liebmann, S. Mansfield, G. Han, J. Culp, J. Hibbeler, R. Tsai, Reducing DfM to practice: the lithography manufacturability assessor, in *Proceedings of the SPIE Advanced Lithography* (2006), pp. 786–798
10. H.D. Ceniceros, G.H. Fredrickson, Numerical solution of polymer self-consistent field theory. Multiscale Model. Simul. **2**(3), 452–474 (2004)
11. N. Laachi, K.T. Delaney, B. Kim, S. Hur, R. Bristol, D. Shykind, C.J. Weinheimer, G.H. Fredrickson, Self-consistent field theory investigation of directed self-assembly in cylindrical confinement. J. Polym. Sci. Part B Polym. Phys. **53**(2), 142–153 (2015)
12. Opencores, http://www.opencores.org/
13. ITC99, http://www.cerc.utexas.edu/itc99-benchmarks/
14. Synopsys, IC Compiler User Guide (2008)
15. S. Shim, W. Chung, Y. Shin, Placement optimization for MP-DSAL compliant layout, in *Proceedings of the International Conference on IC Design and Technology (ICICDT)* (2016), pp. 1–4

Chapter 5
Redundant Via Insertion for DSAL

In DSAL, vias that are physically close are clustered and patterned together through a guide pattern (GP) [1, 2]. A large and complex GP is not allowed to form because it is likely to cause a pattern failure on a wafer. This chapter addresses redundant via insertion problem for DSAL. The goal is to maximally insert redundant vias while vias (both original and redundant) are clustered to form only desirable GPs. The problem can be formulated as finding maximum independent set (MIS) of a conflict graph. Experiments demonstrate that 13% more redundant vias are inserted compared to simple-minded approach, in which a basic insertion with no consideration of DSAL is followed by removal of redundant vias that cause undesirable GPs. DSA defect probability of via cluster is addressed in order to quantitatively define which GPs are allowed during the redundant via insertion process.

5.1 Introduction

Redundant via insertion is a standard practice to preserve the connectivity that via provides under potential via pattern failure. It has become more important as optical lithography approaches its resolution limit to pattern a fine feature [3]. Redundant via is placed at minimum distance from original via as shown in Fig. 5.1, and a pair of original and redundant via is often called a double via. The electrical connection that a double via supports can be kept even though one of its via fails. A few researches [4–6, 9] have studied redundant via insertion problem.

This chapter addresses redundant via insertion for DSAL [8]. The key in this problem is a consideration for via cluster. Figure 5.1a illustrates one example of redundant via insertion, in which undesirable large via cluster forms. Figure 5.1b shows alternative insertion with formation of only good clusters. The redundant via insertion for DSAL is to maximally insert redundant vias while adjacent vias do

S. Shim and Y. Shin, *Physical Design and Mask Synthesis for Directed Self-Assembly Lithography*, NanoScience and Technology,
https://doi.org/10.1007/978-3-319-76294-4_5

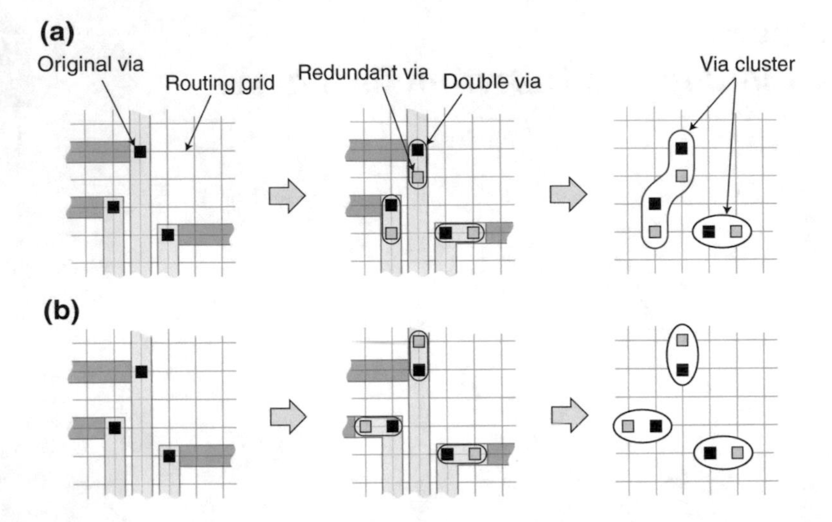

Fig. 5.1 **a** Redundant via insertion at nonideal positions that cause a large via cluster and **b** alternative insertion that is desirable

not form undesirable cluster. The definition of undesirable and desirable via cluster should be made, which is studied in conjunction with DSA defect probability.

5.2 Preliminaries

5.2.1 Defect Probability of Via Cluster

A key component in DSAL redundant via insertion is to define a list of via clusters that are allowed. Consider a via cluster consisting of four vias as shown in Fig. 5.2a. Defect probability of open failure D_{open} is specified for each via, and that of short failure D_{short} is specified for each space between two adjacent vias (see Sect. 2.2). Assume that the two vias (see dotted box in Fig. 5.2b) comprise a double via. Short defect between them does not cause an electrical failure, so their D_{short} is ignored; smaller D_{open} value of the two vias (4.8) is regarded as D_{open} of double via. The maximum value (8.6) in the remaining defect probabilities in turn becomes the defect probability of the via cluster.

It is not practical to calculating defect probability of a via cluster every time it forms during redundant via insertion. Fortunately, a list of candidate clusters that may appear can be identified [8], and the defect probability of each cluster can be calculated beforehand. This is possible because popular gridded design rule (GDR) [10] limits via position, and defect probability is invariant as long as GP image shape is the same for via cluster.

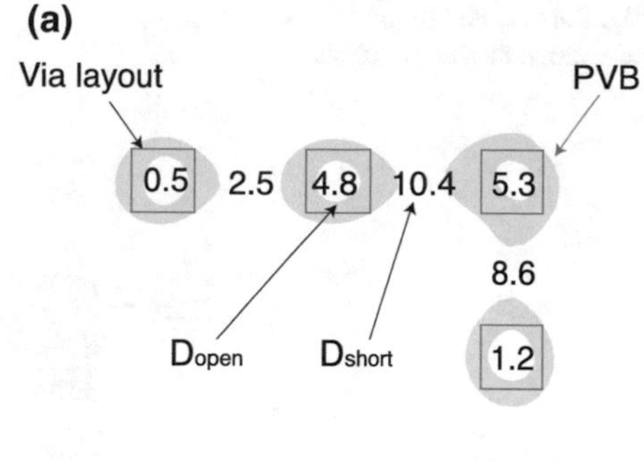

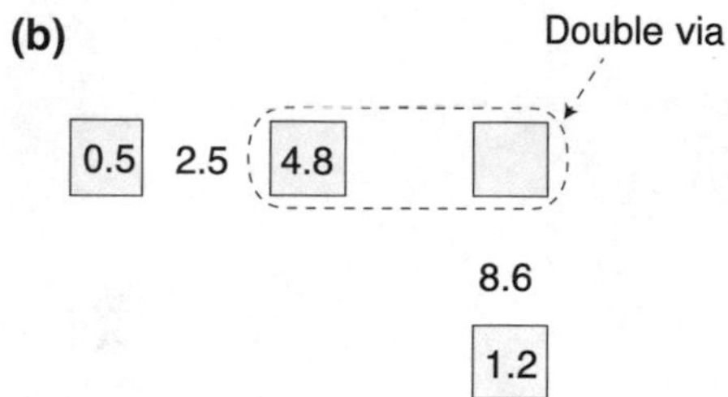

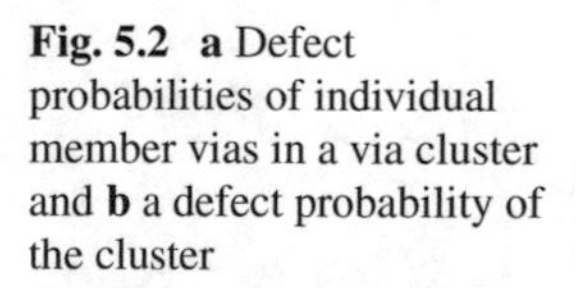
Fig. 5.2 **a** Defect probabilities of individual member vias in a via cluster and **b** a defect probability of the cluster

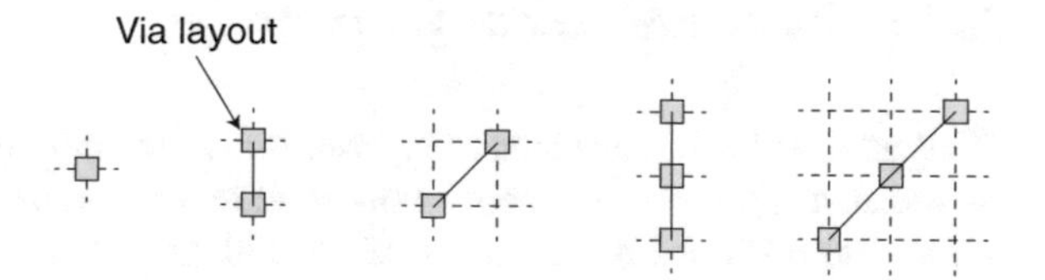

Fig. 5.3 Manufacturable via clusters

Defect probability of via cluster is demonstrated in [8] using 10-nm synthetic via layouts of a few circuits from OpenCores [12] and ITC'99 benchmarks [13]. Logic synthesis, placement, and routing are performed with 15-nm NanGate library [14]; Via1 layouts are appropriately shrunk so that they can follow GDR of 10-nm technology [11], in which via size is 25 nm by 25 nm with minimum via pitch of 50 nm. Redundant vias are inserted with basic insertion method [4] that does not account for DSAL. Any vias that are one grid apart (either in orthogonal or diagonal direction) are clustered, which leads to 48 different cluster structures with four or less vias; clusters containing more than four vias have very high defect probability. Calculating defect probability for the 48 cluster structures takes about 60 h. As cluster size increases (i.e., as cluster includes more vias), defect probability also increases. Five clusters being with zero defect probability shown in Fig. 5.3 are manufacturable. If 5% defect probability is allowed for manufacturing, 11 clusters turn out to be manufacturable.

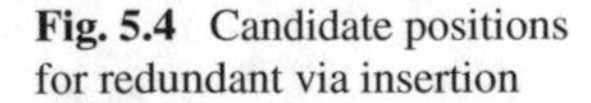
Fig. 5.4 Candidate positions for redundant via insertion

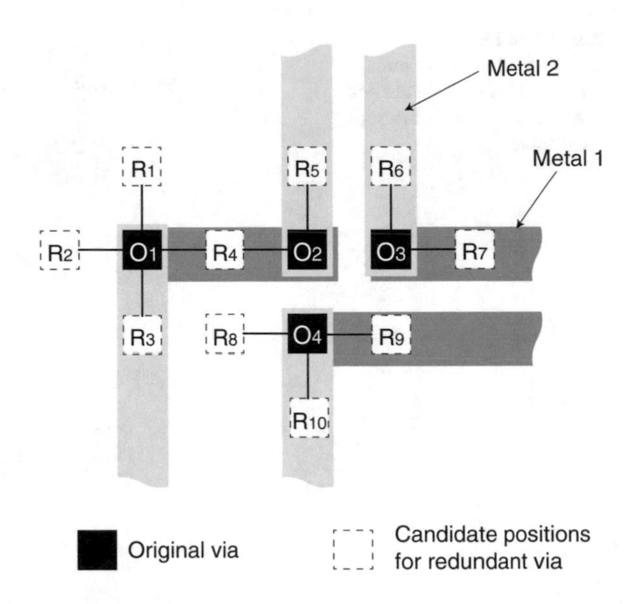

5.2.2 Basic Redundant Via Insertion

The goal of basic redundant via insertion is to maximally insert redundant vias while all design rules and constraints are honored. Figure 5.4 shows an example. All candidate positions of redundant via are identified for each original via, e.g., R_1, R_2, R_3, and R_4 for an original via O_1.[1] Note that R_4 is a position of redundant via for O_1 or O_2, but not for both at the same time; R_9 is not available for O_3 due to electrical short. This basic redundant via insertion problem can be formulated as finding maximum independent set (MIS) [4–6] of an input conflict graph, in which a vertex corresponds to each candidate position of redundant via and an edge is inserted between two nodes if they cannot be inserted at the same time.

5.3 DSAL Redundant Via Insertion Algorithm

The goal of redundant via insertion for DSAL is to maximally insert redundant vias while all design rules are honored and all via clusters are manufacturable, which is obviously more complex than the basic redundant via insertion problem. This problem has been addressed in [7, 8]. DSAL redundant via insertion can also be modeled as a conflict graph, in which a vertex of conflict graph corresponds to a candidate position of redundant via and an edge models a conflict between two

[1]Once redundant via is inserted, metal wires of the same net are locally modified so that the redundant via can be connected. For example, if R_1 is selected, metal 1 and 2 are extended so that both O_1 and R_1 are connected.

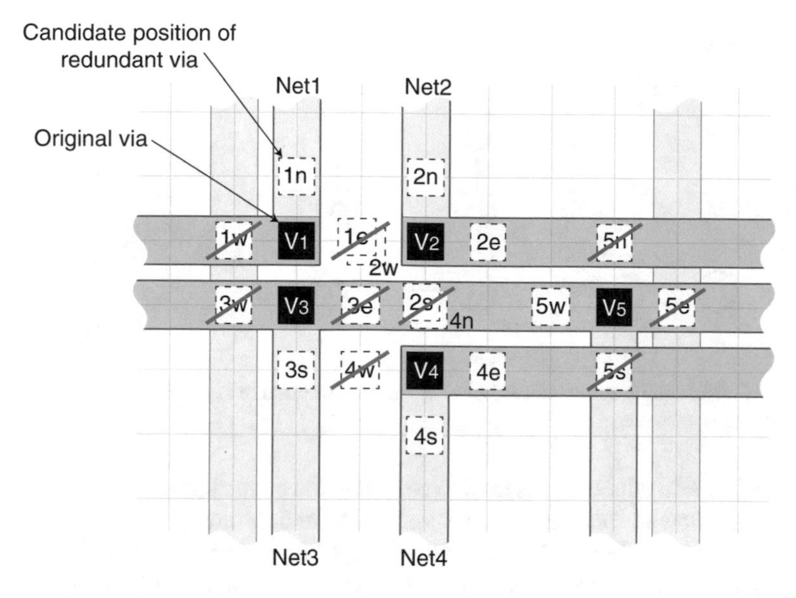

Fig. 5.5 Original vias ($V_1, \ldots, V_5$) and candidate positions of redundant vias

candidate positions. A key is to drop some vertices and add some extra edges so that unmanufacturable via clusters never form in final MIS solution.

5.3.1 *Graph Modeling*

Consider Fig. 5.5, in which five original vias ($V_1, V_2, \ldots, V_5$) are shown. Each original via can have at most one redundant via inserted at one of north, east, west, and south side with minimum pitch distance away. Some guidelines should be followed to determine valid positions for redundant via candidates.

- A redundant via cannot be inserted at position occupied by another original via. For instance, V_3 is not a valid position for V_1's redundant via.
- A redundant via can be located only on the same net. For instance, a redundant via of V_2, which is on Net 2, is not allowed at $2s$, which is on Net 3. Similar reasoning is applied to $1w$, $3w$, $4n$, $5n$, $5e$, and $5s$.
- If grouped redundant via is groped with adjacent original vias into unmanufacturable via cluster, that position is invalid. For example, if a redundant via is introduced at $1e$, it forms an unmanufacturable via cluster with V_1, V_2, and V_3. Similarly, redundant via is not allowed at $2w$, $3e$, and $4w$.

Seven candidate positions are identified for the example in Fig. 5.5, and each position corresponds to a vertex in the conflict graph shown in Fig. 5.6.

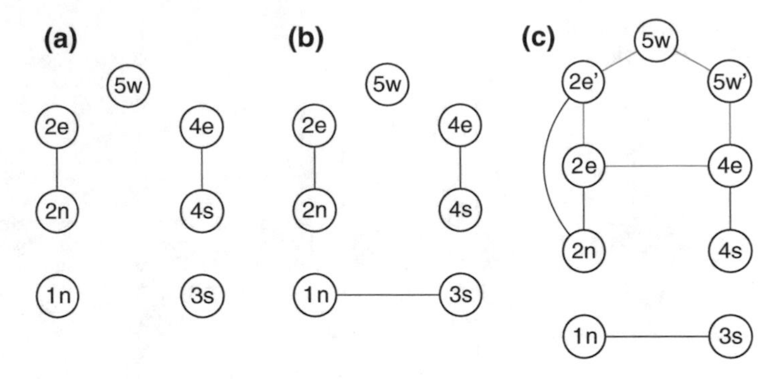

Fig. 5.6 Conflict graph construction for redundant via insertion in Fig. 5.5

Some relevant edges are inserted to complete the conflict graph. Only one redundant via is allowed for each original via, so $2n$ and $2e$ are connected by an edge (Fig. 5.6a) because V_2's redundant via will not be located at two positions at the same time. Similarly, $4e$ and $4s$ are connected. Two vertices will be connected if they correspond to the same position (see, for example, $1e$ and $2w$ even though they are not vertices of conflict graph).

An edge is also introduced if two redundant vias are involved in thc same unmanufacturable via cluster. For example, if redundant vias are inserted at $1n$ and $3s$, four vias ($1n$, V_1, V_3, and $3s$) form unmanufacturable via cluster; $1n$ and $3s$ are connected to reflect this conflict as shown in Fig. 5.6b.

Consider new $2e$, $4e$, and $5w$. If redundant vias are inserted at all three positions, a big via cluster (V_2, $2e$, V_4, $4e$, V_5, and $5w$) forms which is unmanufacturable. However, redundant vias at any two positions do not cause unmanufacturable via cluster. For example, if $4e$, V_5, and $5w$ are big but it is manufacturable. This is because each of a pair of V_4 and $4e$ and a pair of V_5 and $5w$ is a double via, so only possibility of defect is short between $4e$ and $5w$ and open failure in each of four vias, whose probability turns out to be all 0. This situation (three-way conflict) is handled in the conflict graph as follows. Two vertices (say $2e$ and $5w$) are randomly picked and their copies ($2e'$ and $5w'$) are introduced. Relevant edges are attached: an edge between $2e'$ and $2n$ (to duplicate the edge between $2e$ and $2n$), an edge between $2e'$ and $2e$, and an edge between $5w$ and $5w'$ (see Fig. 5.6c). Finally, additional edges are inserted to form a cycle through $2e$, $2e'$, $5w$, $5w'$, and $4e$; note that three vertices cannot be selected from this cycle as an independent set, so three-way conflict is resolved through standard MIS. There is no need to consider a conflict that involves four or more vertices, since a via cluster with such number of vias is not manufacturable.

5.3.2 *Heuristic Insertion Algorithm*

For a conflict graph, there may be several independent sets of the same size. Let $s_{k(i)}$ be the ith independent set whose size is k and S_k be a superset of the independent sets

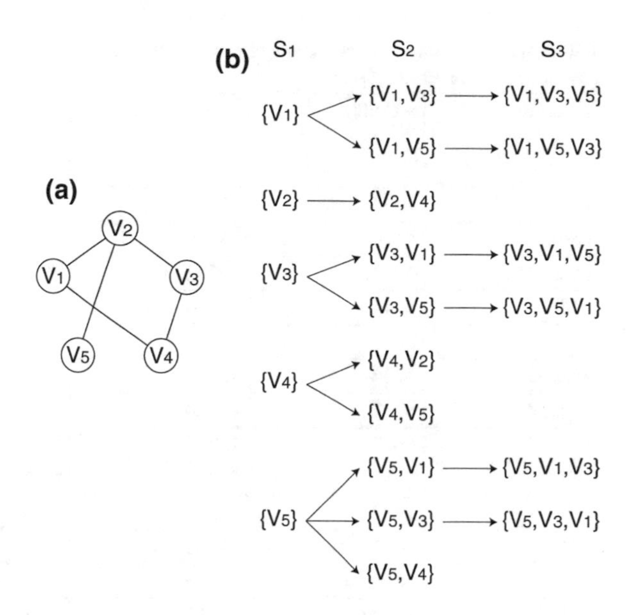

Fig. 5.7 **a** An example of conflict graph and **b** list of independent sets: S_1, S_2, and S_3 corresponds to the superset of the independent sets with size 1, 2, and 3 respectively

of size k. The smallest independent sets of size 1 are first considered. In Fig. 5.7a, for instance, each vertex is an independent set itself (see S_1 in Fig. 5.7b). S_2 is then identified in such a way that for each $s_{1(i)}$, vertices that do not connect with $s_{1(i)}$ are identified and each of them is added to $s_{1(i)}$ yielding an independent set of size 2. Regarding $s_{1(1)} = \{v_1\}$, for instance, v_1 is not connecting to v_3 and v_5, thus two independent sets of size 2 $\{v_1, v_3\}$ and $\{v_1, v_5\}$ are identified. This process is repeated for other independent sets of size 1. Larger independent set is repeatedly identified in similar way until no element can be included in the current size of independent set. In this example, independent sets of size 3 is the largest one (see S_3).

5.4 Experiments

The redundant via insertion algorithm of Sect. 5.3 is tested using a few circuits from OpenCores [12] and ITC'99 benchmarks [13]; they are listed in Table 5.1 with the number of cells in column 2 to indicate the size of each circuit and the number of vias in column 3. Each circuit is synthesized using 15-nm NanGate library [14] while GDR (gridded design rule) [10] is assumed. Metal layers up to M4 are used and placement density is set to 70%. Design rules are intentionally modified so that the post-routing layout is free from via clusters with nonzero defect probability. The layout is then appropriately shrunk for smaller virtual 10-nm technology. Via size is set to 25 nm by 25 nm with minimum pitch of 50 nm.

A basic redundant via insertion method [4] is applied, in which redundant via is inserted without consideration of DSAL. The percentage of vias that have redundant

Table 5.1 Percentage of redundant vias and percentage of via clusters with defects after basic redundant via insertion [4]

Circuits	# Cells	# Vias	Basic RV insertion	
			%RVs	%Clusters with defects
spi	1427	11714	89.3	14.2
tv80	4510	33310	88.3	14.0
mem_ctrl	4968	37541	88.9	13.0
b15	5044	36924	90.0	14.3
b14_1	5256	34687	90.3	13.3
s35932	5766	40877	93.4	11.9
s38584	6382	44683	91.5	13.1
s38417	6609	45838	93.2	12.6
ac97	7058	52720	91.6	12.7
usb_func	7222	63264	83.3	13.1
b14	7421	49630	89.0	14.0
b20_1	10567	69865	90.8	13.3
b21_1	10917	71561	90.1	13.5
aes_cipher	12302	92570	87.3	13.4
b17	15448	110787	89.9	14.0
b20	15608	104402	90.1	13.4
b21	15692	105602	89.4	13.6
b22	23301	155624	90.1	14.1
b18	33509	250170	89.1	13.7
Average			89.8	13.4

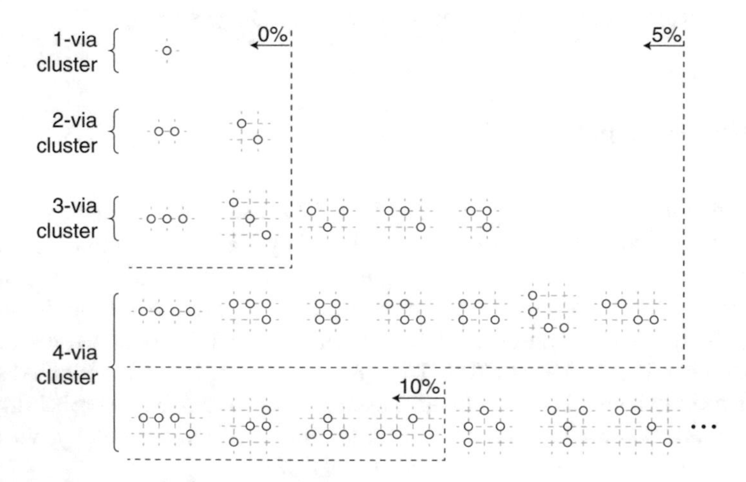

Fig. 5.8 Via clusters of various defect probability

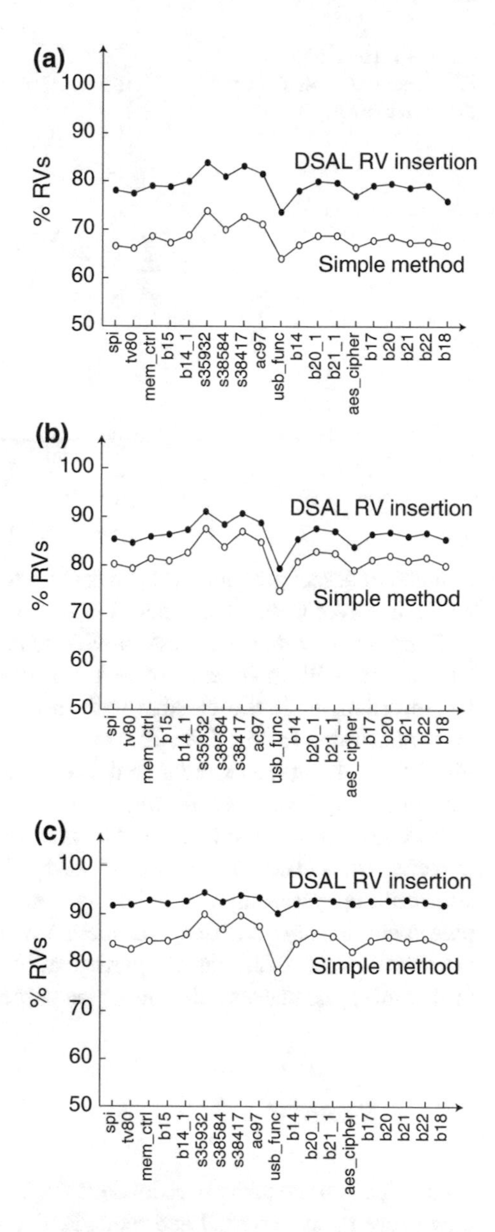

Fig. 5.9 Percentage of redundant vias after simple method and DSAL RV insertion are applied: **a** when no clusters of nonzero defect probability are allowed, **b** when clusters with defect probability less than 5% are allowed, and **c** when clusters with defect probability less than 10% are allowed

vias (percentage of redundant vias for short) is indicated in column 4; average is about 90%. The percentage of via clusters being with nonzero defect probability is shown in the last column with average of 13.4%. In via clusters of nonzero defect probability, all redundant vias are now removed. This method is named "Simple" in Fig. 5.9a. Percentage of redundant vias drops to 75% on average, about 15% reduction from fourth column of Table 5.1. The method of Sect. 5.3 is named "DSAL RV insertion" in Fig. 5.9a. It yields on average of 88% of redundant vias without any via clusters

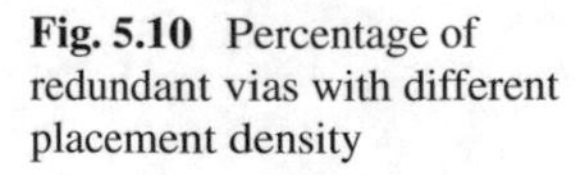
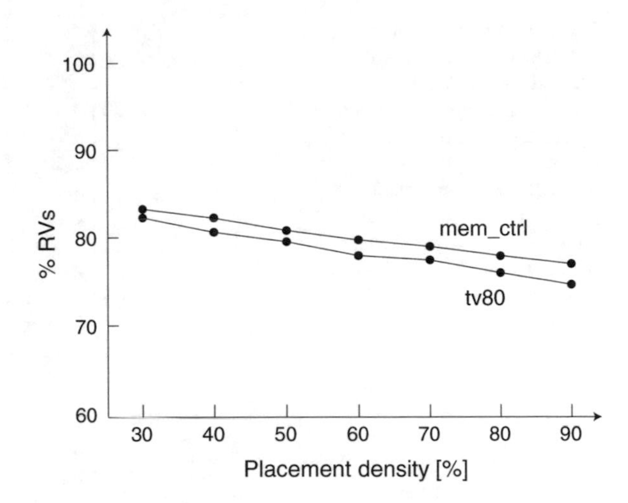

Fig. 5.10 Percentage of redundant vias with different placement density

of nonzero defect probability; 88% is close to 90% from basic insertion in Table 5.1, which however does not consider DSAL.

Suppose now that via clusters with some defect probability are also allowed.[2] Via clusters with their range of defect probability are illustrated in Fig. 5.8. DSAL RV insertion while via clusters with less than 5% defect probability are allowed is demonstrated in Fig. 5.9b; if 10% is used instead of 5%, the result is shown in Fig. 5.9c. Percentage of redundant vias increases in both insertions because more variety of via clusters are allowed.

Two test circuits are taken and their layout is obtained while placement density is varied from 30% up to 90%. For each layout, "DSAL RV insertion" is applied and resulting percentage of redundant vias is obtained. As shown in Fig. 5.10, as placement density increases, vias are more densely placed, which makes redundant via insertion more difficult (in particular, to avoid via clusters with nonzero defect probability) and causes a decline in the percentage of redundant vias.

5.5 Summary

A care needs to be taken in redundant via (RV) insertion for DSAL because one RV may cause many (original and redundant) vias to be clustered and to form a GP of nonzero defect probability. Nevertheless, the problem can still be formulated as a conflict graph and be solved through MIS, as in standard insertion problem. DSA defect probability is important to define which via clusters should be allowed and which should not.

---

[2]Remember that defect probability calculation is somewhat conservative, so clusters with very low defect probability may not actually cause any defects. Precise decision of which probability should be accepted is up to manufacturing details.

References

1. W. Wang, L. Azat, Y. Zou, T. Coskun, A full-chip DSA correction framework, in *Proceedings of the SPIE Advanced Lithography* (2014), pp. 1–11
2. L. Azat, G. Garner, M. Preil, G. Schmid, W. Wang, J. Xu, Y. Zou, Computational simulations and parametric studies for directed self-assembly process development and solution of the inverse directed self-assembly problem. Jpn. J. Appl. Phys. **53**(6), 06JC01–8 (2014)
3. J. Gyvez, Yield modeling and BEOL fundamentals, in *Proceedings of the International Workshop on System-Level Interconnect Prediction* (2001), pp. 135–163
4. K. Lee, T. Wang, Post-routing redundant via insertion for yield/reliability improvement, in *Proceedings of the Asia South Pacific Design Automation Conference* (2006), pp. 303–308
5. C. Pan, Y. Lee, Redundant via insertion under timing constraints, in *Proceedings of the International Symposium on Quality Electronic Design* (2011), pp. 1–7
6. J.-T. Yan, Z.-W. Chen, B.-Y. Chiang, Y.-M. Lee, Timing-constrained yield-driven redundant via insertion, in *Proceedings of the IEEE Asia Pacific Conference on Circuits and Systems* (2008), pp. 1688–1691
7. S. Fang, Y. Hong, Y. Lu, Simultaneous guiding template optimization and redundant via insertion for directed self-assembly, in *Proceedings of the International Conference on Computer-Aided Design* (2015), pp. 410–417
8. W. Chung, S. Shim, Y. Shin, Redundant via insertion in directed self-assembly lithography, in *Proceeding of the Design, Automation and Test in Europe Conference and Exhibition* (2016), pp. 55–60
9. J. Pak, Y. Bei, D.Z. Pan, Electromigration-aware redundant via insertion, in *Proceedings of the Asia South Pacific Design Automation Conference* (2015), pp. 544–549
10. M. Smayling, V. Axelrad, 32 nm and below logic patterning using optimized illumination and double patterning, in *Proceedings of the SPIE Advanced Lithography* (2009), pp. 1–10
11. H. Yi, X. Bao, R. Tiberio, P. Wong, Design strategy of small topographical guiding templates for sub-15 nm integrated circuits contact hole patterns using block copolymer directed self-assembly, in *Proceedings of the SPIE Advanced Lithography* (2013), pp. 1–9
12. Opencores, http://www.opencores.org/
13. ITC99, http://www.cerc.utexas.edu/itc99-benchmarks/
14. Nangate 15 nm open cell library, http://www.nangate.com/

Chapter 6
Redundant Via Insertion for MP-DSAL

In MP-DSAL, vias that are physically close are clustered to form a GP and patterned together through DSAL process; in addition, GPs that are too close are created on a wafer using different masks through MP. It should be very careful to insert redundant vias in MP-DSAL, since some redundant vias may cause large and complex via clusters, whose GPs is likely to cause DSA defect; some other redundant vias may cause MP coloring conflict.

This chapter addresses redundant via insertion for MP-DSAL, which aims to insert maximum number of redundant vias while all GPs are manufacturable and no MP coloring conflict occur. The problem can be formulated as ILP, which is used as a reference for comparison to a graph-based heuristic algorithm.

6.1 Introduction

In MP-DSAL, vias that are closely located are clustered, and a GP image surrounding the cluster is then synthesized. If two GPs are too close beyond optical resolution limit, their mask images are assigned to different masks, which then go through MP process to print the final GPs on a wafer. GPs are then filled with BCPs, which are arranged due to attractive and repulsive forces between polymers and GP wall; one type of polymer is etched down to the substrate leaving final vias.

In the basic DSAL (without MP), via clustering is uniquely determined for a given via layout [1, 2]. However, a large and complex GP with many member vias is not manufacturable because of its high DSA defect probability [3, 4]. This can be alleviated in MP-DSAL by dividing such large GPs into smaller thus manufacturable ones,

S. Shim and Y. Shin, *Physical Design and Mask Synthesis for Directed Self-Assembly Lithography*, NanoScience and Technology,
https://doi.org/10.1007/978-3-319-76294-4_6

which are then assigned to different masks. In this process, various configurations of via clustering and mask assignment are possible, so the corresponding problem is not simple. The problem, called MP-DSAL decomposition, is shown to be NP-complete, and some heuristic algorithms have been proposed [5–8].

The basic redundant via insertion aims to insert maximum number of redundant vias [9–12] while all design rules and constraints are honored. In MP-DSAL, redundant via should be inserted with careful considerations of via clustering and mask assignment. Let us consider a simple example shown in Fig. 6.1. In (a), a large cluster is resulted in due to redundant vias, so the cluster is split into two smaller ones. However, MP coloring conflict occurs due to small distance between clusters, if two masks are only allowed. In (b), on the other hand, different redundant vias yield three small clusters, which can be assigned to two masks without MP coloring conflict.

This chapter introduces the state of the art of redundant via insertion for MP-DSAL, whose key components are summarized as follows:

- DSA defect probability, which is a probability that a via cluster causes patterning failure during DSAL process. It is used to identify manufacturable via clusters during redundant via insertion.
- ILP formulation for redundant via insertion problem for MP-DSAL. It serves as a foundation for developing heuristic algorithm and is also used as a reference of comparison.
- Heuristic algorithm to solve the redundant via insertion problems for MP-DSAL for a practical size of circuit.

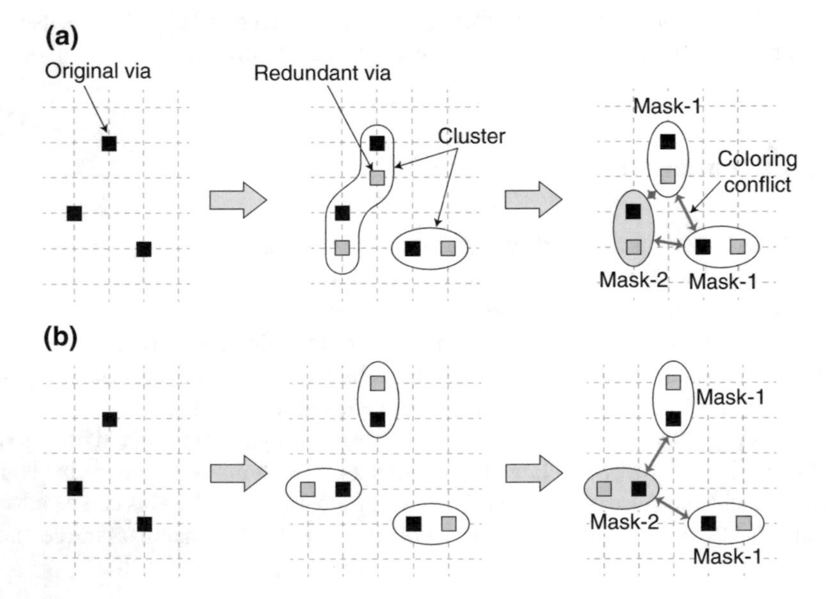

Fig. 6.1 **a** A redundant via insertion causing MP coloring conflict and **b** better insertion with successful clustering and mask assignment

6.2 Simultaneous Optimization of Redundant Via and Via Cluster

6.2.1 ILP Formulation

Redundant via insertion for MP-DSAL aims to maximally insert redundant vias in such a way that all GPs are manufacturable and are successfully assigned to masks without MP coloring conflicts for a given number of masks.

A via layout is first partitioned such that ILP can be formulated independently in each partition. All candidate positions of redundant via insertion are identified for each original via. Two vias (or their respective potential redundant vias) that are p_{litho} or more distance apart can be considered independently during MP-DSAL decomposition (refer to Sect. 4.2). Given via layout is, in turn, partitioned by repeating this check for all via pairs in the layout.

For each partition, all manufacturable via clusters are discovered. For a simple example in Fig. 6.2a, 9 manufacturable clusters are listed as shown in Fig. 6.2b. The manufacturable clusters are then inputted to ILP formulation.

Table 6.1 summarizes the notations for the ILP formulation. The goal is now to determine the values of x_i, y_i, and m_{ik} with objective of maximizing the sum of x_i:

$$\text{Maximize} \sum x_i, \quad \forall r_i \tag{6.1}$$

subject to:

$$y_i + y_j + M_{ij} + C_{ij} \leq 3, \quad \forall\, c_i \text{ and } c_j\ (i \neq j) \tag{6.2}$$

$$\sum_{\forall\, c_i \in Co_k} y_i = 1, \quad \forall\, o_k \tag{6.3}$$

$$\sum_{\forall\, c_i \in Cr_k} x_i \leq 1, \quad \forall\, r_k \tag{6.4}$$

$$\sum_{\forall\, c_i \in CRo_k} x_i \leq 1, \quad \forall\, o_k \tag{6.5}$$

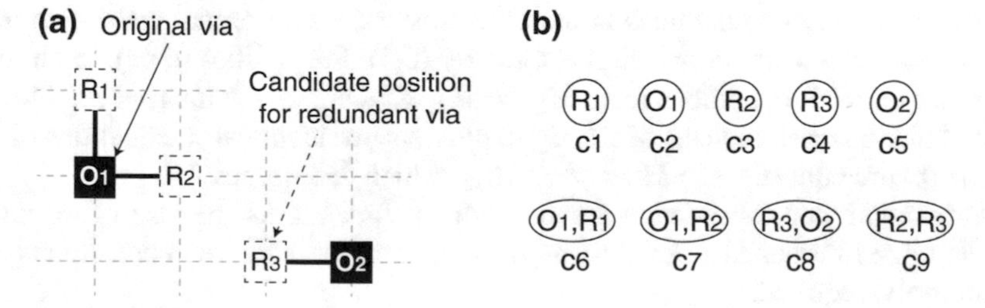

Fig. 6.2 **a** An example via layout with candidate positions of redundant vias and **b** a list of manufacturable via clusters

Table 6.1 Notations of ILP formulation

o_i	i-th original via
r_i	i-th redundant via
Ro_i	A set of redundant vias for o_i
c_i	i-th via cluster
Co_i	A set of clusters containing o_i
Cr_i	A set of clusters containing r_i
CRo_i	A set of clusters that contain redundant vias $\in Ro_i$
C_{ij}	1 if c_i and c_j are in MP coloring conflict, 0 otherwise
M_{ij}	1 if c_i and c_j are assigned to the same mask, 0 otherwise
x_i	1 if r_i is inserted, 0 otherwise
y_i	1 if c_i is selected as a cluster, 0 otherwise
m_ik	k-th bit of the mask index of c_i

$$M_{ij1} \geq 1 - m_{i1} - m_{j1}, \quad \forall\, c_i \text{ and } c_j \ (i \neq j) \tag{6.6a}$$
$$M_{ij1} \leq 1 - m_{i1} + m_{j1}, \quad \forall\, c_i \text{ and } c_j \ (i \neq j) \tag{6.6b}$$
$$M_{ij1} \leq 1 + m_{i1} - m_{j1}, \quad \forall\, c_i \text{ and } c_j \ (i \neq j) \tag{6.6c}$$
$$M_{ij1} \geq m_{i1} + m_{j1} - 1, \quad \forall\, c_i \text{ and } c_j \ (i \neq j) \tag{6.6d}$$
$$M_{ij2} \geq 1 - m_{i2} - m_{j2}, \quad \forall\, c_i \text{ and } c_j \ (i \neq j) \tag{6.6e}$$
$$M_{ij2} \leq 1 - m_{i2} + m_{j2}, \quad \forall\, c_i \text{ and } c_j \ (i \neq j) \tag{6.6f}$$
$$M_{ij2} \leq 1 + m_{i2} - m_{j2}, \quad \forall\, c_i \text{ and } c_j \ (i \neq j) \tag{6.6g}$$
$$M_{ij2} \geq m_{i2} + m_{j2} - 1, \quad \forall\, c_i \text{ and } c_j \ (i \neq j) \tag{6.6h}$$

$$2M_{ij} \leq M_{ij1} + M_{ij2}, \quad \forall\, c_i \text{ and } c_j \ (i \neq j) \tag{6.7a}$$
$$M_{ij} \geq M_{ij1} + M_{ij2} - 1. \quad \forall\, c_i \text{ and } c_j \ (i \neq j) \tag{6.7b}$$

Inequalities (6.2)–(6.5) correspond to the following constraints: (6.2) is mathematically equivalent to $x_i x_j M_{ij} C_{ij} = 0$, which prohibits coloring conflict; (6.3) constrains every original vias to be picked only once; (6.4) imposes that one redundant via should not be included in multiple clusters; each original via should have one or less redundant via, which is limited by (6.5). Inequalities (6.6a)–(6.6h) correspond to XNOR operations (e.g., $M_{ij1} = m_{i1}$ XNOR m_{j1}), which set M_{ijk} to 1 if the kth bits of mask indexes of i and jth clusters are identical. Inequalities (6.7a) and (6.7b) are equivalent to $M_{ij} = M_{ij1} M_{ij2}$, which determines M_{ij}.

Aforementioned formulation assumes MP of four masks. In case of assuming two and three masks, i.e., double- and triple patterning, additional constraints are, respectively, required:

$$m_{i2} = 0, \qquad \forall\, c_i \tag{6.8}$$

and

$$m_{i1} + m_{i2} \leq 1. \qquad \forall\, c_i \tag{6.9}$$

6.2.2 Graph-Based Heuristic

Two conflict graphs are constructed for a heuristic algorithm. The algorithm consists of two steps: (1) redundant via insertion and via clustering, which aim to find a set of manufacturable clusters with maximum number of redundant vias and least MP coloring conflicts; (2) MP mask assignment of the via clusters.

Graph Modeling

For each via partition, all manufacturable via clusters are discovered as shown in Fig. 6.2. A cluster containing both original- and redundant-via corresponds to a vertex; for instance, c_6, c_7, and c_8 in Fig. 6.2b correspond to v_2, v_3, and v_6, respectively, as shown in Fig. 6.3. A cluster composed of a redundant via is paired with an original via that the redundant via is associated with, and the pair corresponds to a vertex; for example, c_1 in Fig. 6.2b is paired with c_2, and v_1 corresponds to the cluster pair in Fig. 6.3; similarly, c_3 and c_4 correspond to v_4 and v_8, respectively. Note that since c_9 is composed of R_2 and R_3, which are associated with two original vias O_1 and O_2, c_9 is paired with each of c_2 and c_5, respectively (see v_5 and v_7 in Fig. 6.3). Each vertex has a weight value, which is the number of redundant vias in the vertex and is indicated in a bracket in Fig. 6.3.

After all vertices are generated, two conflict graphs are constructed by inserting some edges:

- **Via Conflict Graph** (Fig. 6.3a): If two vertices contain the same via, they have an edge in-between (see red edge between v_5 and v_8 due to R_3); this corresponds to inequalities (6.3) and (6.4) in the ILP formulation. If two vertices contain redundant vias that are originated from the same original via, they also have an edge in-between (see black edge between v_2 and v_7 due to R_1 and R_2, which are originated from O_1); this corresponds to inequality (6.5) in the ILP formulation. Note that multiple edges between two vertices are merged.
- **Coloring Conflict Graph** (Fig. 6.3b): If any two clusters from two vertices are in MP coloring conflict, the two vertices have an edge in-between. Note that if two vertices have an edge in-between in the via conflict graph, they do not have an edge in the coloring conflict graph.

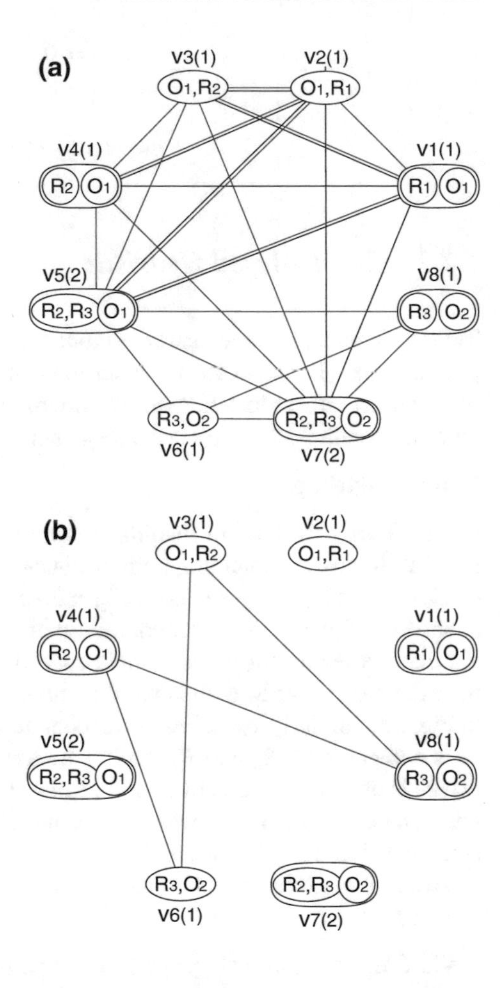

Fig. 6.3 **a** A via conflict graph and **b** coloring conflict graph for the heuristic approach. Each vertex has a weight value (in a bracket), which indicates the number of redundant vias in the corresponding cluster

Algorithm

In the first step, the algorithm attempts to select a set of vertices that are not connected in the via conflict graph, while sum of their weights is maximized. This can be formulated as the maximum weight independent set (MWIS) problem, which is solved using a heuristic similar to [13].

For each vertex, its neighbors are identified in the via conflict graph. The score of the current vertex is then determined as weight values of the neighboring vertices subtracted from the weight value of the current vertex; for instance, the score of v_6 is thus -4 in Fig. 6.3a. After calculating scores of all vertices, the highest scored vertex is picked as a member of MWIS; if multiple vertices have the highest score, the vertex that has the least number of edges in the coloring conflict graph is picked; this makes mask assignment problem easier in the next step. The picked vertex and its neighbors are then removed from both graphs. This process is repeated for every remaining vertex until no vertex remains.

In Fig. 6.3, for example, v_6 and v_8 have the highest score (−4), but v_6 is picked as a member of MWIS because v_8 consists of two clusters in MP coloring conflict. v_6 and its neighbors (such as v_5, v_7, and v_8) are removed and the scores are then re-calculated. The remaining four vertices (such as v_1, v_2, v_3, and v_4) have the same score, but v_2 is finally picked since v_1 is composed of clusters in MP coloring conflict and v_3 and v_4 have edges in the coloring conflict graph. Therefore, v_2 and v_6 turns out to be the output of the first step.

In the next step for mask assignment, a coloring conflict graph is newly constructed using all vertices from the first step. A vertex containing multiple clusters is split into multiple vertices, each of which corresponds to one cluster. If two clusters in MP coloring conflict, the corresponding vertices have an edge in-between. This graph is then inputted to a commercial mask assignment tool [14], which outputs a set of vertices that are successfully assigned to given masks and another set of vertices that are failed in mask assignment. In the example of Fig. 6.3, v_2 and v_6 are inputted to this step, and they can be assigned to the same mask because they have no edge in-between.

6.3 Experiments

The ILP formulation and heuristic algorithm are implemented using Tcl script and GUROBI [15], which is used as an ILP solver; Calibre tool [14] is used to solve MP mask assignment problem. An intuitive method (sequential) serves as a reference for comparison; in Sequential, a standard method of redundant vias insertion [9] is employed, and MCM-based heuristic [6] is applied for via clustering and MP mask assignment.

A few test circuits from OpenCores [16] and ITC99 benchmarks [17] are synthesized using 15-nm NanGate library [18]. Via-1 layout is then appropriately shrunk and modified so that the layout can follow virtual 7-nm design rule, in which via size is assumed to be 15 nm by 15 nm, and metal track pitch is set to 35 nm. p_{dsa} and p_{litho} are also assumed to be 50 and 100 nm, respectively, which are compared with some typical via pitch as shown in Fig. 6.4. The number of vias ranges from 10 k (spi) to 450 k (b19) as shown in the second column of Table 6.2. The percentage of vias that have redundant vias (percentage of redundant vias for short) using a standard

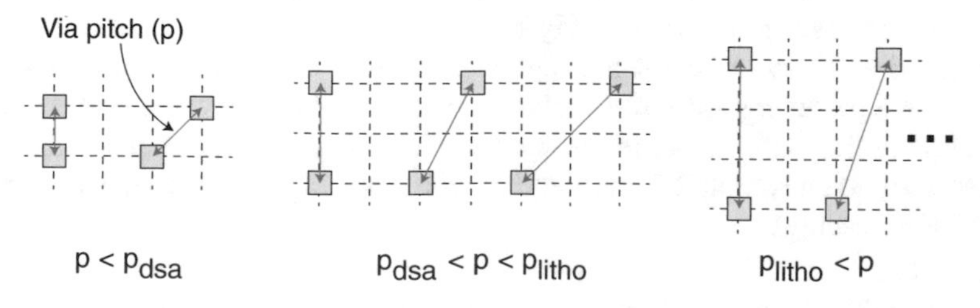

Fig. 6.4 Some typical via pairs and their pitches, which are compared to p_{dsa} and p_{litho}

Table 6.2 Comparisons of three methods in triple patterning

Circuits	#Vias	Standard RV insertion [9]		RV insertion for MP-DSAL					
				Sequential		ILP		Heuristic	
		#RVs (%)	#Defects	Time (s)	#RVs (%)	Time (s)	#RVs (%)	Time (s)	#RVs (%)
spi	11175	92.7	7434	15	81.7	5592	90.8	9	89.7
tv80	31757	91.3	21167	33	77.9	16889	89.4	26	89.0
b15	35227	92.5	23554	31	78.2	18254	90.8	30	90.3
b14	47184	94.1	31377	38	79.1	21174	92.4	34	92.1
usb_func	60743	93.1	39962	61	79.7	29501	90.7	49	89.7
b21	100730	92.8	67463	95	78.9	–	–	80	90.4
b22	147936	92.5	99474	132	78.2	–	–	116	90.0
b18	238789	91.5	160507	227	77.9	–	–	207	88.1
Ethernet	273460	93.8	159740	255	78.4	–	–	267	92.2
b19	468316	91.7	314603	469	78.2	–	–	407	90.1
Average		92.6			78.8				90.2

method [9] is shown in column 3. The percentage is about 93% on average, but a large number of unmanufacturable clusters with nonzero defect probability remain as indicated in the fourth column.

Assessment of Heuristic Algorithm

Sequential, ILP, and heuristic methods are compared in terms of runtime and the percentage of vias that receive redundant vias in Table 6.2. In sequential, only 78.8% vias receive redundant vias on average across test circuits. Heuristic, on the other hand, inserts on average of 90.2% redundant vias, which is quite comparable to standard insertion (column 3) even though all clusters are manufacturable. For the first five circuits, where ILP can complete within 10 h, heuristic shows very comparable result to ILP, which demonstrates the efficiency of heuristic.

Number of MP Masks

In Table 6.2, three masks (triple patterning or TP) have been assumed. Assuming two masks (double patterning or DP), more coloring conflicts will occur and less number of redundant vias will be allowed to be inserted due to less degree of freedom in mask assignment. The three insertion methods with DP are applied to the first four test circuits in Table 6.2. The percentage of vias that receive redundant vias, on average of the four circuits, is compared in Fig. 6.5.

In all three methods, the percentage significantly decreases in case of DP. In addition, the percentage becomes similar over circuits, because those three methods can insert no (or only a few) redundant vias for many large partitions[1]; for those groups, the percentages in ILP and Heuristic are much higher than that in Sequential when TP is assumed.

[1] Remind that via partitions are the input of each method as an input as presented in Sect. 6.2.1.

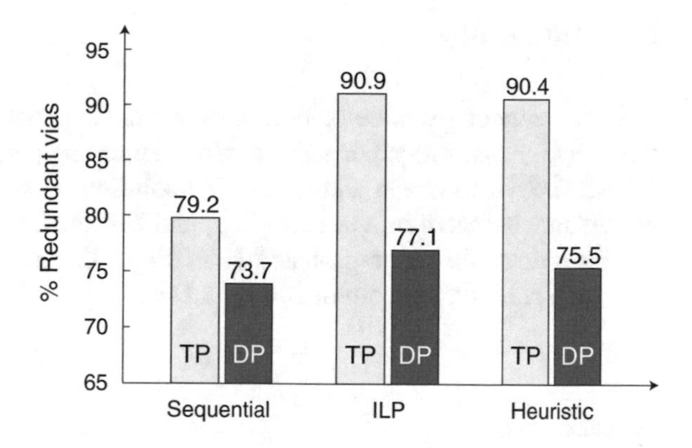

Fig. 6.5 Percentage of vias that receives redundant vias with triple patterning (TP) and double patterning (DP)

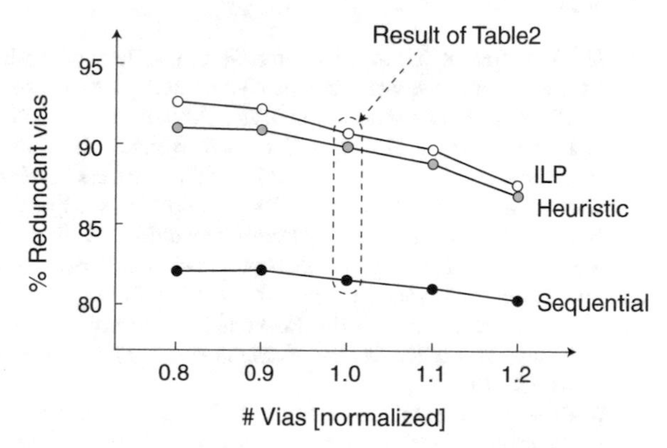

Fig. 6.6 Percentage of vias that receive redundant vias with varying via density in test layout spi

Via Density

As via density increases, via clusters are likely to be physically close. So, more MP coloring conflicts may occur, which makes the problem more difficult. To prove this conjecture, for one test layout (spi), some vias are randomly picked and are gradually removed for generating synthetic layouts of lower via density (x-axis value smaller than 1.0 in Fig. 6.6); similarly, some additional vias are gradually introduced at random locations to create a synthetic layout of higher density (x-axis value larger than 1.0).

Percentage of vias that receive redundant vias of the three insertion methods is compared in Fig. 6.6. Note that the percentage when normalized via density (x-axis) is 1.0 corresponds to Table 6.2 results for spi. As via density increases, the percentage decreases as expected; most part of redundant via loss occurs to large via partitions that contain 8 or more original vias. The percentage of the three methods gets similar as via density increases, which is understandable consequence because the portion of via partitions receiving a similar percentage of redundant vias increases in the three methods.

6.4 Summary

In 7-nm technology node or below, even DSAL is not enough and it is expected that DSAL is used together with multiple patterning (MP). Redundant via insertion for MP-DSAL has been addressed. The solution needs to properly coordinate the redundant via insertion, via clustering, and MP mask assignment. ILP formulation has been presented; a graph-based heuristic algorithm has also been described for the corresponding problem of a practical size.

References

1. H. Yi, X. Bao, R. Tiberio, P. Wong, Design strategy of small topographical guiding templates for sub-15 nm integrated circuits contact hole patterns using block copolymer directed self-assembly, in *Proceedings of the SPIE Advanced Lithography*, pp. 1–9 (2013)
2. L. Azat, G. Grant, P. Moshe, S. Gerard, W. Wong, J. Xu, Y. Zou, Computational simulations and parametric studies for directed self-assembly process development and solution of the inverse directed self-assembly problem. Jpn. J. Appl. Phys. **53**(6S), 1–8 (2014)
3. S. Shim, W. Chung, Y. Shin, Defect probability of directed self-assembly lithography: fast identification and post-placement optimization, in *Proceedings of the International Conference on Computer Aided Design*, pp. 404–409 (2015)
4. W. Chung, S. Shim, Y. Shin, Redundant via insertion in directed self-assembly lithography, in *Proceedings of the Design, Automation and Test in Europe Conference and Exhibition*, pp. 55–60 (2016)
5. Y. Badr, A. Torres, Y. Ma, J. Mitra, P. Gupta, Incorporating DSA in multipatterning semiconductor manufacturing technologies, in *Proceedings of the SPIE Advanced Lithography*, pp. 1–8 (2015)
6. Y. Badr, A. Torres, P. Gupta, Mask assignment and synthesis of DSA-MP hybrid lithography for sub-7 nm contacts/vias, in *Proceedings of the Design Automation Conference*, pp. 70:1–70:6 (2015)
7. S. Shim, W. Chung, Y. Shin, Redundant via insertion for multiple-patterning directed-self-assembly lithography, in *Proceedings of the Design Automation Conference*, pp. 41:1–41:6 (2016)
8. J. Gyvez, Yield modeling and BEOL fundamentals, in *Proceedings of the International Workshop on System-Level Interconnect Prediction*, pp. 135–163 (2001)
9. K. Lee, T. Wang, Post-routing redundant via insertion for yield/reliability improvement, in *Proceedings of the Asia South Pacific Design Automation Conference*, pp. 303–308 (2006)
10. C. Pan, Y. Lee, Redundant via insertion under timing constraints, in *Proceedings of the International Symposium on Quality Electronic Design*, pp. 1–7 (2011)
11. J. Yan, Z. Chen, B. Chiang, Y. Lee, Timing-constrained yield-driven redundant via insertion, in *Proceedings of the IEEE Asia Pacific Conference on Circuits and Systems*, pp. 1688–1691 (2008)
12. S. Fang, Y. Hong, Y. Lu, Simultaneous guiding template optimization and redundant via insertion for directed self-assembly, in *Proceedings of the International Conference on Computer Aided Design*, pp. 410–417 (2015)
13. S. Sakai, M. Togasaki, K. Tamazaki, A note on greedy algorithm for the maximum weighted independent set problem. Discret. Appl. Math. **126**(2), 313–322 (2003)

14. *Calibre Multi-Patterning Manual*, Mentor Graphics, Jan. 2013
15. Gurobi Optimization, Inc., Gurobi optimizer reference manual, http://www.gurobi.com/
16. Opencores, http://www.opencores.org/
17. ITC99, http://www.cerc.utexas.edu/itc99-benchmarks/
18. Nangate 15 nm open cell library, http://www.nangate.com/

Part II
Mask Synthesis and Optimizations

Chapter 7
DSAL Mask Synthesis

In DSAL, a mask contains the image of guide patterns (GPs) not the image of final contacts or vias. Thus, a wafer receives GP patterns after optical lithography is applied to the mask. It then goes through a DSA process, and contacts are finally patterned. Mask design for DSAL, which is in fact the opposite of the above mentioned process, consists of two key steps, inverse DSA and inverse lithography. In inverse DSA, GPs are progressively refined until they are estimated to produce contacts that are close to target ones as much as possible. For this purpose, GP is defined as a function of a few geometry parameters, and the sensitivity of contact to each parameter is repeatedly calculated to guide how much GP should be refined. In inverse lithography, mask is progressively refined until it is believed to produce target GPs. Mask is defined by a grid of pixel values (rather than by a set of edges) and their gradient guides the direction that the mask should be refined. There are typically too many pixels for gradient calculation; the method to approximate calculation is described in this chapter. Inverse DSA and inverse lithography are extended to handle process variations. The basic inverse lithography is modified so that the resulting mask becomes less sensitive to lithography variations; basic inverse DSA is modified so that it provides the way this sensitivity can be checked.

7.1 Introduction

Figure 7.1 compares how mask is synthesized in optical lithography and in DSAL. In optical lithography shown in (a), a lithography image (litho image for short) is obtained for a given layout; it then goes through optical proximity correction (OPC) to yield a final mask image. Mask design in DSAL is more complex as illustrated in (b). A DSA image is extracted from a layout; note that DSA image is an expected result of DSA process (see Figs. 1.8 and 2.2) while litho image results from lithography

S. Shim and Y. Shin, *Physical Design and Mask Synthesis for Directed Self-Assembly Lithography*, NanoScience and Technology,
https://doi.org/10.1007/978-3-319-76294-4_7

process, which is why they are named differently even though the same process p1 is applied to extract them. Nearby contacts in DSA image are clustered [1–3] and a guide pattern image (GP image) is compiled for each cluster. Only the contours in GP image are taken (now called GP litho image), from which the final mask image is designed.

This chapter covers three important components of DSAL mask synthesis.

- **Inverse DSA**: A key process in Fig. 7.1b is inverse DSA, labeled p3. There have been a few studies for the opposite process, e.g., form various shapes of GP images on a wafer and check how contacts are formed for each GP [2–5], and verify whether a given GP image causes any unexpected contact shapes [6, 7]. But inverse DSA has been studied less extensively. An analytical solution to inverse DSA does not exist because a GP image is not uniquely determined for a DSA image. An iterative solution has been studied [8] and is explained in Sect. 7.2, in which a GP image is progressively refined while the resulting DSA image, obtained through DSA simulation, is assessed against a target DSA image. A GP image is defined as a function of a few geometry parameters; it is refined by using sensitivity matrix, which contains the extent of how sensitive the DSA image is to each parameter change.
- **Inverse lithography**: Another important process in Fig. 7.1b is p2* named inverse lithography. This is similar to p2 of Fig. 7.1a. A prime difference exists in how mask image is defined. In conventional OPC, a mask image is defined by a set of edges. Even in some DSAL techniques, edge-based mask image has been used to simplify mask design process [3]. But GP mask image should be more accurately defined to accommodate complex curve shapes that GP litho image contains. In the approach in Sect. 7.3, a mask image is defined as a collection of pixels. It is progressively refined using gradient descent method, and resulting litho image is assessed for each refinement. Gradient calculation is approximated without resorting to explicit lithography simulation, which speeds up the process.
- Inverse DSA and inverse lithography can be extended to account for DSA- and lithography variations (Sect. 7.4). Two problems are addressed. For a given error tolerance of DSA image, the error tolerance of GP image is derived, which is in fact the outer and inner boundaries within which the actual GP image has to reside; this is solved by applying inverse DSA multiple times. In the second problem, the basic inverse lithography is modified in a way that final mask yields GP image that always resides within the error tolerance of GP image even though lithography settings are varied.

7.2 Inverse DSA

An algorithm to solve inverse DSA (process p3 of Fig. 7.1b) is shown in Fig. 7.2. It receives one cluster of contacts in DSA image ($\mathcal{D}_{in}$), and returns corresponding GP image. An initial GP image is constructed (L1) by placing circles (each is a concentric circle for a circle in DSA image with radius being set to the length of

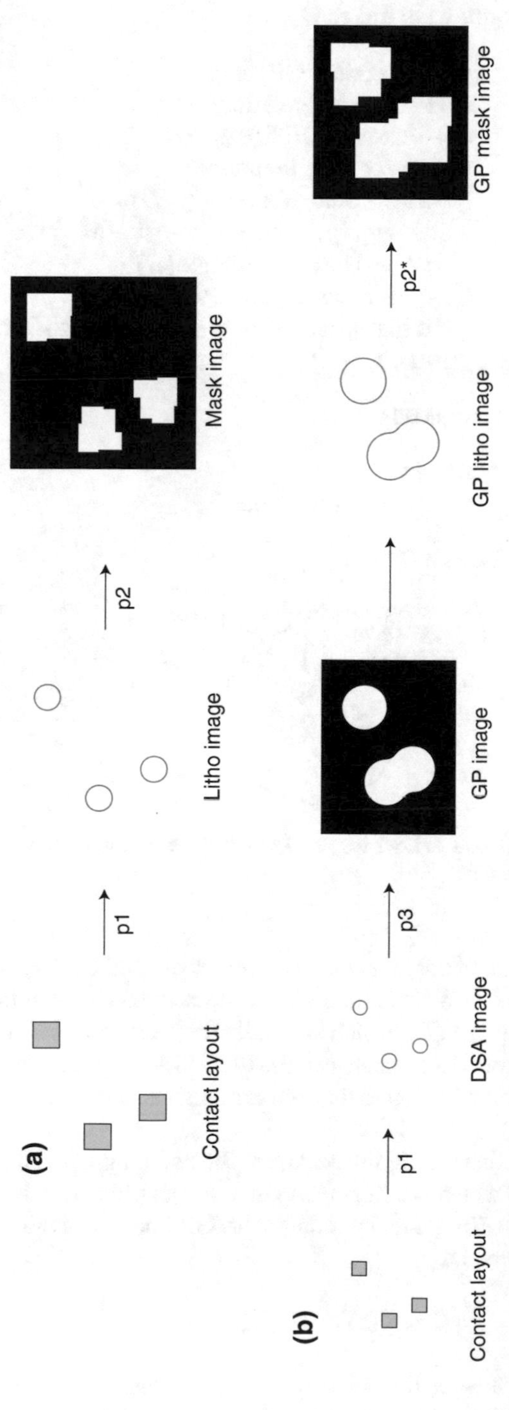

Fig. 7.1 Mask design for **a** optical lithography and **b** DSAL

Input: a DSA image $\mathcal{D}_{in}$ of a cluster of contacts
Output: a GP image $\mathcal{G}$

L1: $\mathcal{G} \leftarrow$ an initial GP image
L2: $\mathcal{D} \leftarrow$ DSA_Simulation($\mathcal{G}$)
L3: $\mathbf{e} \leftarrow$ Measure_EPE($\mathcal{D}_{in}$, $\mathcal{D}$)
L4: **repeat** for max_iterations
L5: $\quad \mathbf{M} \leftarrow$ Calc_Matrix($\mathcal{D}_{in}$, $\mathcal{G}$)
L6: $\quad \mathcal{G} \leftarrow f(\mathbf{g})$, where $\mathbf{g}^{\mathrm{T}} \leftarrow \mathbf{g}^{\mathrm{T}} - \mathbf{M}^{-1} \times \mathbf{e}^{\mathrm{T}}$
L7: $\quad \mathcal{D} \leftarrow$ DSA_Simulation($\mathcal{G}$)
L8: $\quad \mathbf{e} \leftarrow$ Measure_EPE($\mathcal{D}_{in}$, $\mathcal{D}$)
L9: $\quad$ **if** $\max_{\forall i}(|e_i|) \leq$ max_EPE **then** Exit loop
L10: **return** $\mathcal{G}$

Fig. 7.2 Algorithm for inverse DSA

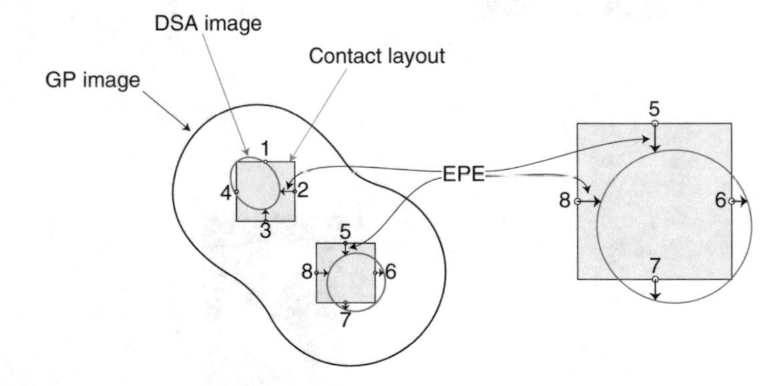

Fig. 7.3 GP image ($\mathcal{G}$) and its DSA image ($\mathcal{D}$); each contact of input DSA image has 4 EPEs at certain measurement points

one BCP string), taking only the boundary arc of merged circles, and smoothing the boundary (see Fig. 7.3). A DSA simulation is performed on the initial GP image (L2) to obtain its DSA image ($\mathcal{D}$), which is usually different from input DSA image. The edge placement errors (EPEs) between the two DSA images are measured at certain points (four for each circle), and they are arranged as a vector $\mathbf{e} = (e_1, e_2, \ldots, e_m)$ (L3; see Fig. 7.3).

The goal is to refine the GP image so that the resulting DSA image resembles the input DSA image (which is a target) as much as possible. This is achieved through iteration (L4 to L9). To guide the refinement, GP image is defined as a function of geometry parameters, i.e.,

$$\mathcal{G} = f(\mathbf{g}) = f(g_1, g_2, g_3, \ldots, g_n). \tag{7.1}$$

The parameters g_i include the radius of each constituent circle, the center-to-center distance between adjacent circles, the angle between adjacent center-to-center lines,

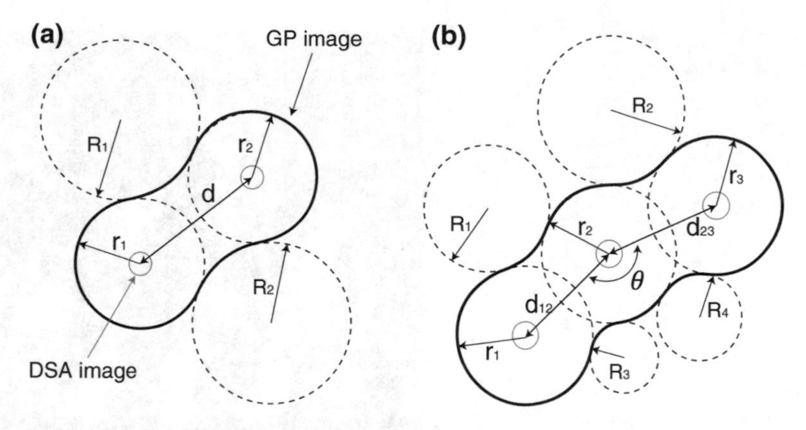

Fig. 7.4 Geometry parameters of GP image

and radius of each circumscribed circle as shown in Fig. 7.4. Each parameter is perturbed to obtain a new GP image, a DSA simulation is performed that returns corresponding DSA image, and its EPEs are assessed; a set of EPE changes allows the introduction of the sensitivity matrix:

$$\mathbf{M} = \begin{bmatrix} \frac{\partial e_1}{\partial g_1} & \frac{\partial e_1}{\partial g_2} & \cdots & \frac{\partial e_1}{\partial g_n} \\ \frac{\partial e_2}{\partial g_1} & \frac{\partial e_2}{\partial g_2} & \cdots & \frac{\partial e_2}{\partial g_n} \\ \cdots & \cdots & \cdots & \cdots \\ \frac{\partial e_m}{\partial g_1} & \frac{\partial e_m}{\partial g_2} & \cdots & \frac{\partial e_m}{\partial g_n} \end{bmatrix}. \tag{7.2}$$

It can be shown that the following holds:

$$\mathbf{M} \times \Delta\mathbf{g}^{\mathrm{T}} = \Delta\mathbf{e}^{\mathrm{T}}, \tag{7.3}$$

where $\Delta\mathbf{e}$ is EPE variations due to some extents of the parameter changes ($\Delta\mathbf{g}$). Rearranging (7.3) yields

$$\Delta\mathbf{g}^{\mathrm{T}} = \mathbf{M}^{-1} \times \mathbf{e}^{\mathrm{T}}, \tag{7.4}$$

where $\Delta\mathbf{g}$ now indicates how much each parameter should be adjusted to compensate current EPEs (L6) to obtain a new GP image.[1] DSA simulation and EPE measurement then follow (L7–L8). If the largest EPE does not exceed a certain threshold ($\max_{\forall i}(|e_i|) \leq$ max_EPE), iteration stops; otherwise iteration continues until user-defined maximum iterations are reached.

[1]If $\mathbf{M}$ is not square, $\mathbf{g}^{\mathrm{T}} - (\mathbf{M}^{\mathrm{T}} \times \mathbf{M})^{-1} \times \mathbf{M}^{\mathrm{T}} \times \mathbf{e}^{\mathrm{T}}$ replaces $\mathbf{g}^{\mathrm{T}}$ or singular value decomposition (SVD) is applied in L6.

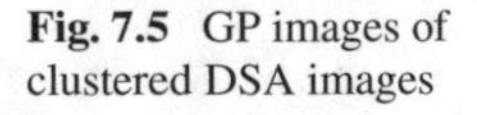

Fig. 7.5 GP images of clustered DSA images

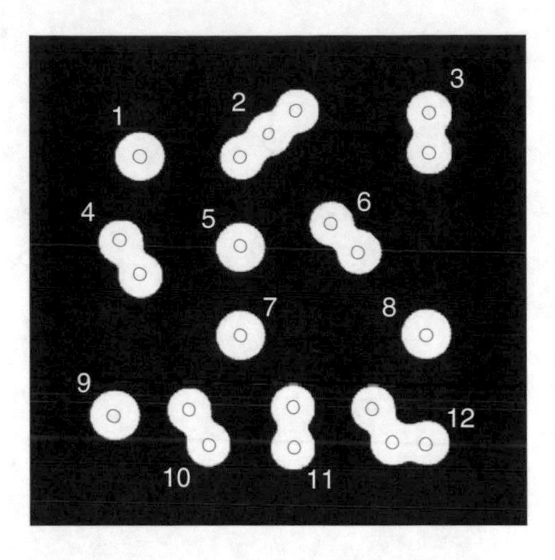

The algorithm in Fig. 7.2 synthesizes a GP image for one cluster, so it has to be applied to all clusters (of contacts and vias) in a layout. Fortunately, DSA process is localized within a GP image,[2] and so the same DSA image of a cluster should correspond to the same GP image. As an example of Fig. 7.5, a GP image is synthesized and shared for clusters of {1, 5, 7, 8, 9}; the same applies to {3, 4, 6, 10, 11}, {2}, and {12}; note that {3, 4, 6, 10, 11} are not exactly the same but they share the same geometry parameters. In sample test layout that contains 804 clusters, only 31 (or about 4%) requires inverse DSA.

7.2.1 Numerical Results

The algorithm of inverse DSA is implemented in MATLAB and C++; DSA simulator is based on self-consistent field theory [9, 10]. A few test circuits from Opencores [11] are synthesized by using 15 nm NanGate library [12], and layouts in contact and via layers are scaled down so that the layouts follow the design rules of 10 nm technology. In contact, size is 22 nm and minimum pitch is 45 nm; via has a size of 25 nm with minimum pitch of 50 nm.

The number of geometry parameters determines the number of DSA simulations to calculate the sensitivity matrix (7.2), and a DSA simulation takes more time as the area of GP image increases. Figure 7.6 illustrates the runtime of inverse DSA for various types of clusters. For the same number of contacts (or vias) in a cluster, the runtime varies substantially. Clusters (a) and (b) have smaller number of parameters due to the existence of symmetry, while (c) and (d) have more parameters and take more time for inverse DSA.

[2] It has been reported that DSA process is affected by GP density due to under-filling or over-filling of GP [1]. But, this can be overcome by additional process steps, e.g., DSA planarization [5].

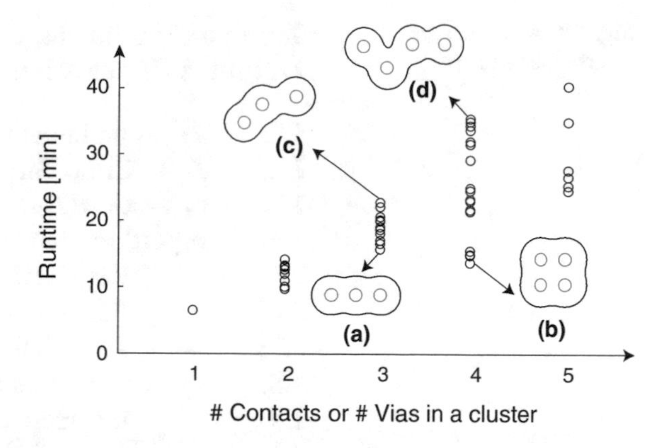

Fig. 7.6 Runtime of inverse DSA for various contact/via clusters

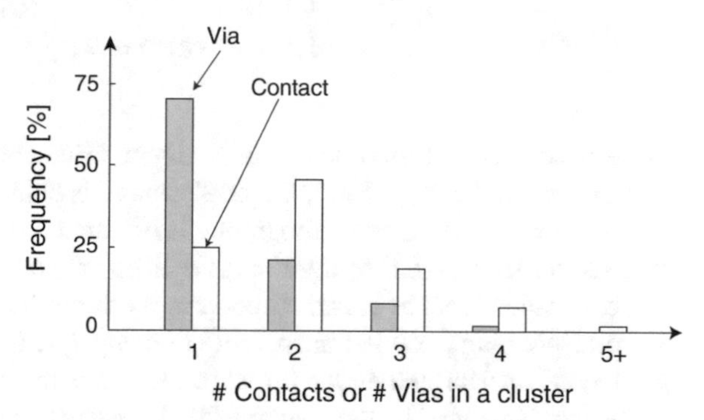

Fig. 7.7 Frequency of clusters with different number of contacts/vias

Figure 7.7 illustrates how the clusters of different number of contacts and vias are distributed for one sample circuit. Generally, clusters have smaller number of vias (shaded bars) because recent physical design tends to avoid routing congestion thereby spreading out vias. In contact layer (white bars), on the other hand, there are many large clusters, which contain two or more contacts. Distribution of contact clusters is more affected by how cells are designed rather than by how cells are placed. This is partly because many cells have whitespace in-between. Note that a few very large clusters containing more than 5 contacts appear in the boundary of abutted cells, and they appear more frequently as placement density increases.

7.3 Inverse Lithography

The input to inverse lithography (process p2* in Fig. 7.1b) is a GP litho image, which is simply a GP contour.[3] In corresponding process p2 of a conventional OPC, mask

[3]In fact, input is the whole GP contours of a layout, but we will consider a single GP contour for simplicity of explanation.

Fig. 7.8 Algorithm for inverse lithography

Input: a GP litho image $\mathcal{L}_{in}$
Output: a GP mask image $\mathcal{M}$

L1: $\mathcal{M} \leftarrow$ an initial GP mask image
L2: $\mathcal{L} \leftarrow$ Litho_Simulation($\mathcal{M}$)
L3: $C \leftarrow$ Cost($\mathcal{L}_{in}$, $\mathcal{L}$)
L4: **repeat** for max_iterations
L5: $\mathcal{M} \leftarrow \mathcal{M} - k\nabla C$
L6: $\mathcal{M} \leftarrow$ Convert $\mathcal{M}$ to a binary mask
L7: $\mathcal{L} \leftarrow$ Litho_Simulation($\mathcal{M}$)
L8: $C \leftarrow$ Cost($\mathcal{L}_{in}$, $\mathcal{L}$)
L9: **if** C increases OR $|\nabla C| \leq \varepsilon$ **then**
L10: Roll back $\mathcal{M}$; exit loop
L11: **return** $\mathcal{M}$

image is constructed by repeated adjustment of edges. In the approach of inverse lithography, on the other hand, GP mask image is obtained by adjusting pixels.

The algorithm of inverse lithography is illustrated in Fig. 7.8. An initial GP mask image is constructed (L1) by digitizing GP litho image, i.e., if a pixel is within the litho image, its value is set to 1, and it becomes 0 otherwise. The value indicates whether the corresponding pixel is transparent (1) or opaque (0) to the light. A lithography simulation is performed on the initial mask image to obtain litho image (L2).

The cost is defined by the sum of EPEs between current litho image ($\mathcal{L}$) and input litho image ($\mathcal{L}_{in}$). A few points on input litho image are sampled to measure EPEs, as shown in Fig. 7.9. An imaginary line is drawn such that it passes the centers of adjacent contacts; the lines that are orthogonal to such line are also drawn from the centers of contacts. EPEs are measured at the points where the lines and input litho image intersect (points at peak); another points are also located at concave (points at valley); some more points are identified (points along arc) between adjacent points at peak and valley in regular interval.

The gradient of cost (∇C) is defined by

$$\nabla C = \left(\frac{\partial C}{\partial g_1}, \frac{\partial C}{\partial g_2}, \frac{\partial C}{\partial g_3}, \ldots, \frac{\partial C}{\partial g_n}\right), \tag{7.5}$$

where parameters (g_i) are now pixel values rather than geometry properties; its calculation is addressed in Sect. 7.3.1. A new mask image is constructed by adjusting each parameter of current mask image in the direction of negative gradient with step size k (L5), which is empirically determined. Some pixel values become nonbinary in this process and so they are digitized (L6), i.e., if pixel value is larger than 1, it is set to 1, and it is set to 0 otherwise. A lithography simulation is then performed to obtain litho image (L7), whose cost is evaluated through EPE calculation (L9). If cost increases or refined mask image yields an insignificant change in litho image

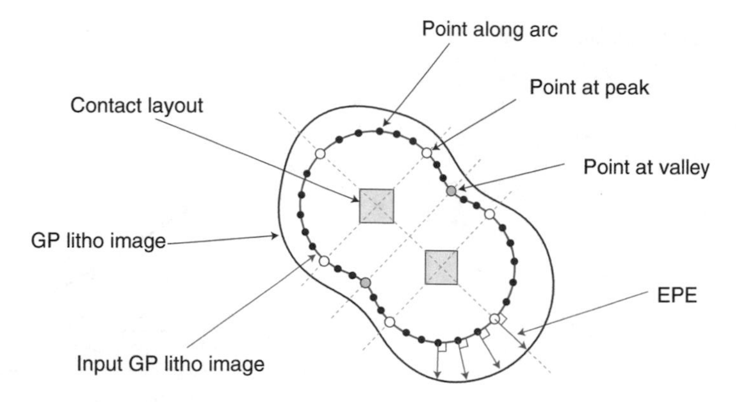

Fig. 7.9 EPE measurement points

($|\nabla C| \leq \varepsilon$), iteration stops with previously refined mask image; otherwise iteration continues until user defined maximum iterations are reached.

7.3.1 Approximation of Cost Gradient

An element of ∇C ($\partial C/\partial g_i$) can be obtained by perturbing one pixel value (g_i), performing lithography simulation, and assessing litho image through EPE calculation. This is computationally too expensive simply because there are many pixels. Instead, the gradient calculation is approximated without explicit lithography simulation.

The light energy exposed on a wafer through mask image ($\mathcal{M}$), called intensity, determines litho image:

$$I(x, y) = |\mathcal{M}(x, y) \otimes H(x, y)|^2, \tag{7.6}$$

where H models optical element including light source and lenses, and $\otimes$ indicates a convolution. Let pixel value at (x_i, y_i) be perturbed; a new mask image can be represented by

$$\mathcal{M}'(x, y) = \mathcal{M}(x, y) + \delta(x - x_i, y - y_i), \tag{7.7}$$

where δ is Dirac delta function. Corresponding intensity is

$$\begin{aligned} I' &= |\mathcal{M}' \otimes H|^2 \\ &= |(\mathcal{M} + \delta(x - x_i, y - y_i)) \otimes H|^2 \\ &= I + |H(x - x_i, y - y_i)|^2 + H^*(x - x_i, y - y_i)(\mathcal{M} \otimes H) \\ &\quad + H(x - x_i, y - y_i)(\mathcal{M} \otimes H)^*, \end{aligned} \tag{7.8}$$

where * denotes conjugate transpose.

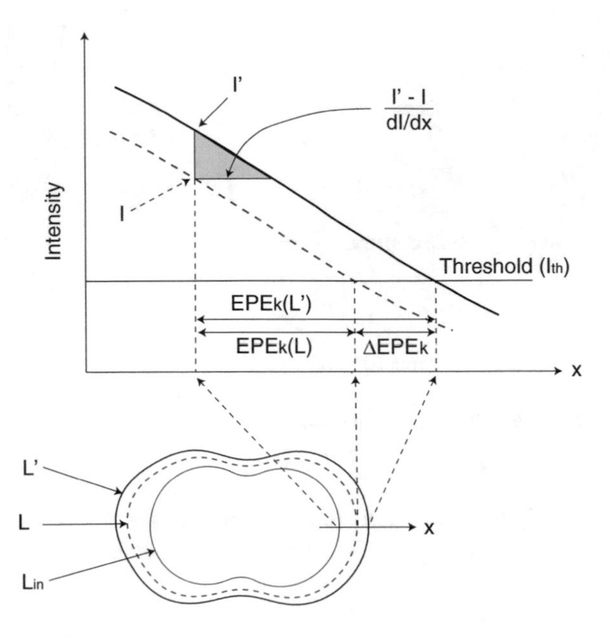

Fig. 7.10 Approximation of ΔEPE_k

An approximation of $\partial C/\partial g_i$ is tried using (7.8). By definition of cost:

$$\partial C/\partial g_i = \sum_k \big(\text{EPE}_k(\mathcal{L}', \mathcal{L}_{in}) - \text{EPE}_k(\mathcal{L}, \mathcal{L}_{in})\big)$$
$$= \sum_k \Delta\text{EPE}_k, \tag{7.9}$$

where $\mathcal{L}'$ is a potential new litho image if pixel value g_i is perturbed, and measurement point is indexed by k. Figure 7.10 plots the two intensities I and I' with x-axis corresponding to the direction of EPE measurement. I_{th} denotes a threshold intensity, i.e. pattern is formed on a wafer unless intensity is smaller than I_{th}. I and I' are assumed to be parallel, which is valid if x is in a small range of value. By referring to Fig. 7.10, it can be verified that EPE_k is equal to the bottom side of the shaded triangle, i.e.,

$$\Delta\text{EPE}_k = \frac{I' - I}{dI(x_k, y_k)/dx}. \tag{7.10}$$

Note that $I' - I$ is readily obtained from (7.8); $\mathcal{M} \otimes H$ is available from lithography simulation (L2 or L7), and $H(x - x_i, y - y_i)$, which is, in fact, $H(x_k - x_i, y_k - y_i)$ in (7.10), is simply obtained from known H.

Table 7.1 Comparison of exact and approximate inverse lithography

Layout	Exact method			Approximate method		
	# Iter	Time (hours)	EPE_{max} (nm)	# Iter	Time (hours)	EPE_{max} (nm)
Via 1	5	1	0.3	11	0.1	0.9
Via 2	5	1.8	0.6	13	0.2	1.1
Via 3	6	4.2	0.6	14	0.5	0.7
Contact 1	7	4.8	0.5	18	0.4	0.8
Contact 2	7	6.8	0.5	17	1.1	0.7
Contact 3	10	16.7	0.7	25	2.9	1.4
Average	6.7	5.9	0.5	16.3	0.9	0.9

7.3.2 Evaluation

The algorithm of inverse lithography is implemented in MATLAB and C++, and is applied to 6 test layouts (contact and via layers of 3 sample circuits). Since the layout is too large (120 μm by 120 μm on average) to handle at the same time, it is divided into smaller tiles of 18 μm by 18 μm regions; the algorithm is then serially applied to each region and aggregated results yield a whole mask image; to handle optical proximity effect in the boundaries of the regions, in fact 20 μm by 20 μm region is considered even though the result is taken from central 18 μm by 18 μm region. ArF immersion lithography process is assumed with 1.35 numerical aperture (NA) and an annular shape of illumination.

Inverse lithography with approximation of cost gradient (named approximate method) is compared to the one that performs explicit lithography simulations (named exact method) for gradient calculations. The maximum number of iterations for all tiles, algorithm runtime (total runtime for all tiles using a single CPU), and maximum EPE value (out of all GPs in the test layout) of final litho image are compared in Table 7.1. For the same circuit, via takes less time than contact, e.g., Via 1 versus Contact 1. This is because vias are usually sparser than contacts, for instance, one tile contains 269 GP contours on average in Contact 1, while Via 1 has just 48 GP contours within one tile. Approximate method requires more iterations due to inaccuracy of gradient computation, which implies that mask refinement is not always performed in ideal direction; runtime nevertheless is reduced by 85% on average thanks to simple approximation, which is not associated with any simulations. Note that runtime for a full-chip layout of 10 mm by 10 mm area is estimated to be about 10 hours when the industrial computing environment is assumed (e.g., distributed processing using 500 cores), which is practically reasonable runtime [13]. Maximum EPE slightly increases, but is still within an acceptable range.[4]

[4]About 1–2% error in GP image incurs about 4–5% error in DSA image. Patterning margin in DSA image is ±10%.

Contact requires more iterations because of worse accuracy from approximation. This is because contacts are usually denser than vias, so GP contours are closer to each other, which increases optical proximity effect and leads to decreased slope of intensity curve and thus smaller value of dI/dx (see Fig. 7.10); delta EPE value in (7.10) increases as a result.

7.4 Mask Design with Process Variations

Both DSA and optical lithography are inherently associated with process variations. The mask design for DSAL (Fig. 7.1b) is studied again while such variations are taken into account.

7.4.1 *Inverse DSA and Inverse Lithography*

Consider inverse DSA, which takes DSA image as an input and yields GP image as an output (Fig. 7.11a). DSA image is typically associated with error tolerance in $\pm 10\%$ range. Such error tolerance is used to alter geometry of DSA image such as circle size and distance between circles, which yields a series of altered DSA images (Fig. 7.11b); inverse DSAs are then applied to obtain corresponding GP images. All contours of GP images are assembled (Fig. 7.11c); the inner and outer boundaries of aggregated contours are defined as the error tolerance of GP image.

Consider now inverse lithography; its basic algorithm in Fig. 7.8 is modified to take account of lithography variations:

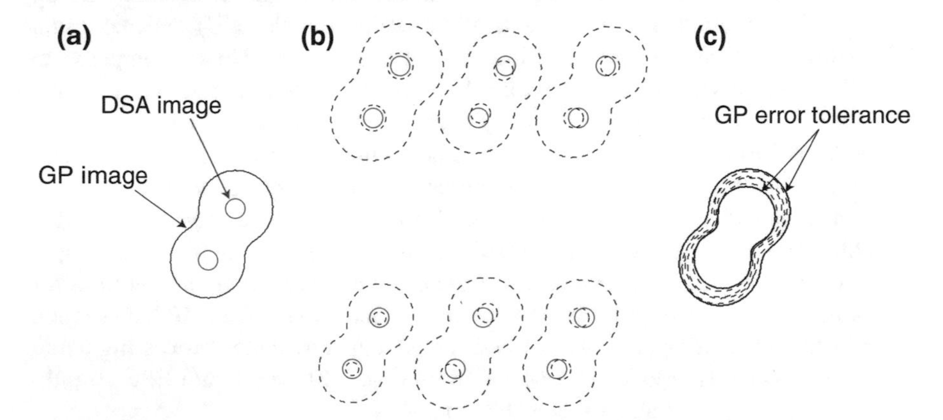

Fig. 7.11 Process to derive GP error tolerance: **a** DSA and GP images without error; **b** DSA and GP images (dotted) with various errors; **c** merged GP images from **b** and GP error tolerance (inner and outer boundaries)

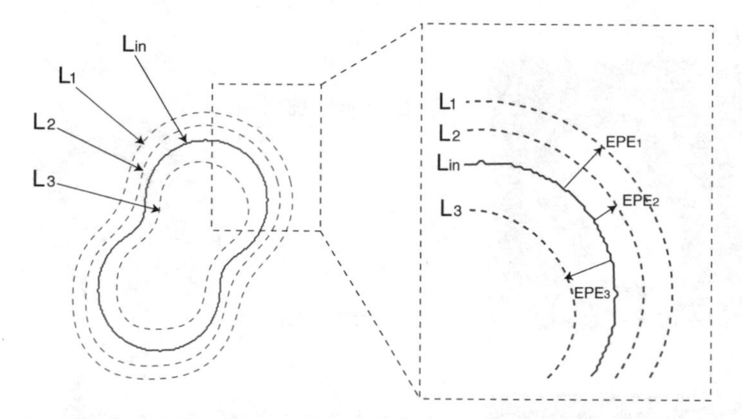

Fig. 7.12 EPE calculation in inverse lithography with lithography variations

- Assist features (AFs) are inserted while an initial mask is synthesized (L1), so that mask is less affected by lithography variations. AFs are also accounted for when the mask is refined (L5). The details of AF insertion are presented in Sect. 7.4.2.
- To calculate cost (L3 and L8), lithography simulations are performed multiple times while lithography settings such as scanner focus, exposure energy, and mask error [14] are varied. The simulations now yield multiple litho images ($\mathcal{L}_1$, $\mathcal{L}_2$, and $\mathcal{L}_3$ in Fig. 7.12); the maximum of all EPE values (at particular measurement point) is considered to be the final EPE value.

Once a final mask is generated, multiple lithography simulations are performed again to check whether all the resulting litho images reside within the error tolerance of GP image.

7.4.2 Insertion of DSA-Aware Assist Feature

AFs are inserted so that the light going through them constructively interferes with the light passing through regular wafer patterns. The amount of light interference can be predicted by an interference map, which is a convolution of input mask image with optical element model $H(x, y)$ in (7.6). The regions of positive value are where the light constructively interferes. AFs are typically inserted into the region where the value is higher than some threshold.

If AF becomes too big during mask refinement process, it may be patterned out on a final wafer, which is not desirable. This can be prevented by restricting the AF size, e.g., keeping the width of its litho image of AF within a certain value, which is set by preliminary test.

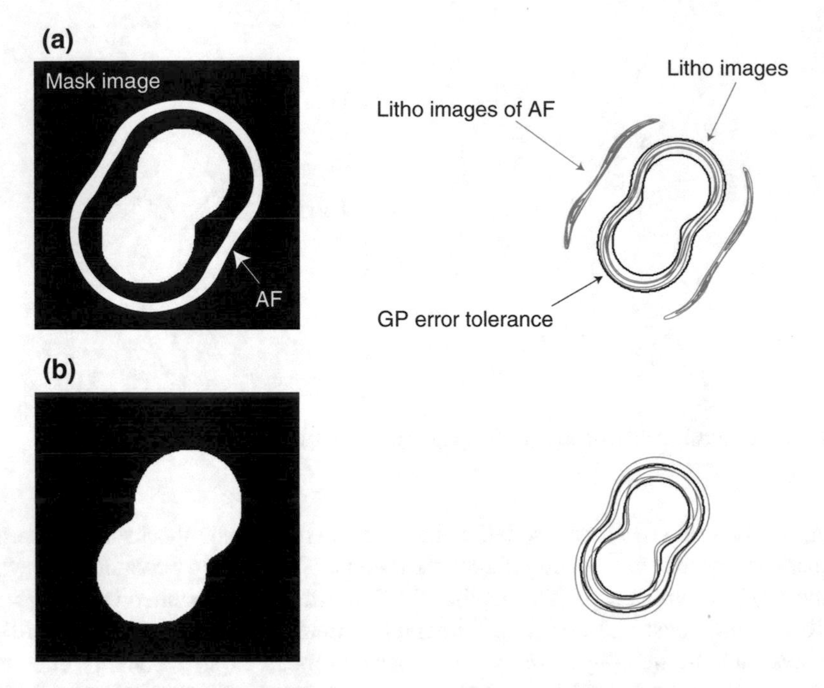

Fig. 7.13 Mask image **a** with AF and **b** without AF; litho images at various lithography settings are compared with GP error tolerance

Figure 7.13a shows a GP mask image with curvilinear AF[5] and its litho images with various lithography settings; GP error tolerance is also shown. The litho images reside within the GP error tolerance, which implies that the variations of DSA image due to the process variations will not exceed the given error tolerance of the contact layout; note that AF also yields litho images, but they are not patterned out on a final wafer. If a mask image does not contain AF, some litho images may reside outside of GP error tolerance as shown in Fig. 7.13b.

7.4.3 Assessment

The new inverse lithography algorithm is assessed against the basic algorithm of Fig. 7.8, and the result is shown in Table 7.2. The same test cases from Table 7.1 are used. Each algorithm is applied to generate a mask image, which is then submitted to lithography simulator while 27 combinations (of scanner focus, exposure energy, and mask error) of lithography setting is tried. The resulting 27 litho images (or 27 GP contours) are checked to see whether they all reside within GP error tolerance; the corresponding GP is marked violated if checking fails. As Table 7.2 indicates,

[5]The curvilinear AF is supported by advanced mask writing technique [15].

Table 7.2 Basic inverse lithography versus modified inverse lithography with lithography variations

Layout	Inverse litho		Inverse litho with litho variations	
	# Violations (%)	Max PVB (nm)	# Violations (%)	Max PVB (nm)
Via 1	7.9	8.3	0.0	3.6
Via 2	6.3	7.2	0.0	3.2
Via 3	6.8	8.8	0.0	3.3
Contact 1	5.3	9.4	0.0	4.4
Contact 2	8.8	9.8	0.0	4.7
Contact 3	4.3	10.5	0.0	5.5
Average	6.6	9.0	0.0	4.1

the new algorithm causes no violations while 6.6% of GPs from the basic algorithm fail the checking.

The maximum distance between the inner and outer contours of 27 litho images is called process variation band (PVB), which indicates the extent of GP sensitivity to process variations. The maximum PVB value out of all GPs is picked and shown under columns "Max PVB". Better result is also observed in this regard (on average of 4.1 nm vs. 9.0 nm).

7.5 Summary

Two key problems of DSAL mask design have been addressed. In inverse DSA, a GP is progressively refined to get its ideal shape. The sensitivity matrix has been introduced to guide GP refinement, in which matrix element represents the sensitivity of member contacts to each geometry parameter of GP. In inverse lithography, a mask is refined to produce target GPs. The refinement is guided by cost gradient while mask is represented by pixels. Gradient calculation has been approximated to reduce large runtime, which is typical in practical layout of huge number of pixels. Inverse DSA and inverse lithography have been extended to handle process variations. In particular, basic inverse lithography algorithm has been modified so that it produces a mask which is less sensitive to lithography variations.

References

1. H. Yi, X. Bao, R. Tiberio, P. Wong, Design strategy of small topographical guiding templates for sub-15nm integrated circuits contact hole patterns using block copolymer directed self-assembly, in *Proceedings of SPIE Advanced Lithography* (2013), pp. 1–9

2. H. Yi, Y. Bao, J. Zhang, C. Bencher, L. Chang, X. Chen, R. Tiberio, J. Conway, H. Dai, Y. Chen, S. Mitra, H.-S.P. Wong, Flexible control of block copolymer directed self-assembly using small, topographical templates: potential lithography solution for integrated circuit contact hole patterning. Adv. Mater. **14**(23), 3107–3114 (2012)
3. L. Azat, G. Grant, P. Moshe, S. Gerard, W. Wong, J. Xu, Y. Zou, Computational simulations and parametric studies for directed self-assembly process development and solution of the inverse directed self-assembly problem. Jpn. J. Appl. Phys. **53**(6S), 1–8 (2014)
4. M. Muramatsu, M. Iwashita, T. Kitano, T. Toshima, M. Somervell, Y. Seino, D. Kawamura, M. Kanno, K. Kobayashi, T. Azuma, Nanopatterning of diblock copolymer directed self-assembly lithography with wet development. J. Micro/Nanolithogr. MEMS MOEMS **11**(3), 1–6 (2012)
5. P. Barros, A. Gharbi, A. Sarrazin, R. Tiron, N. Posseme, S. Barnola, S. Bos, C. Tallaron, G. Claveau, X. Chevalier, M. Argoud, I. Servin, C. Navarro, C. Nicolet, C. Lapeyre, C. Monget, DSA planarization approach to solve pattern density issue, in *Proceedings of SPIE Advanced Lithography* (2015), pp. 1–10
6. Z. Xiao, Y. Du, H. Tian, M. Wong, H. Yi, H. Wong, H. Zhang, Directed self-assembly (DSA) template pattern verification, in *Proceedings of Design Automation Conference* (2014), pp. 1–6
7. S. Shim, Y. Shin, Fast verification of guide patterns for directed self-assembly lithography, IEEE Trans. CAD Integr. Circuits Syst. (accepted for publication)
8. S. Shim, Y. Shin, Mask optimization for directed self-assembly lithography: inverse DSA and inverse lithography, in *Proceedings of Asia South Pacific Design Automation Conference* (2016), pp. 83–88
9. H. Ceniceros, G. Fredrickson, Numerical solution of polymer self-consistent field theory. Multiscale Model. Simul. **2**(3), 452–474 (2004)
10. N. Laachi, K. Delaney, B. Kim, S. Hur, R. Bristol, D. Shykind, C. Weinheimer, G. Fredrickson, Self-consistent field theory investigation of directed self-assembly in cylindrical confinement. J. Polym. Sci. Part B Polym. Phys. **53**(2), 142–153 (2015)
11. Opencores, http://www.opencores.org/
12. Nangate 15nm open cell library, http://www.nangate.com/
13. Samsung Electronics Corp. OPC principal engineer, personal communication, September 2014
14. L. Liebmann, S. Mansfield, G. Han, J. Culp, J. Hibbeler, R. Tsai, Reducing DfM to practice: the lithography manufacturability assessor, in *Proceedings of SPIE Advanced Lithography* (2006), pp. 786–798
15. A. Fujimura, Improvement of mask write time for curvilinear assist feature, in *Proceedings of SPIE Advanced Lithography* (2010), pp. 1–10

Chapter 8
Verification of Guide Patterns

Guide patterns (GPs) are very critical in contacts (and vias) patterning in directed self-assembly lithography (DSAL). Simulations may be used to verify whether each GP will lead to correct patterning of its member contacts, but runtime is excessive. Instead, the shape of GP can be characterized using some geometric parameters. Then, a function for the verification can be constructed to predict whether the required contacts can be obtained by a GP [1, 2]. Specifically, each GP in a test set is represented as a vector in parameter space; DSA simulation is applied to each GP assessing its acceptability, and corresponding vector is marked "good" or "bad" accordingly; the parameter space is deformed in a way that a radial distribution is converted into one in which the good and bad vectors can be successfully separated by a certain hyperplane, which finally becomes the verification function. It is also shown that principal component analysis (PCA) can be applied for reducing the dimensionality of the parameter space, and the characterization of GPs can be generalized to allow different types of GP to be verified in a unified fashion. Such methods are demonstrated in 10 nm technology.

8.1 Introduction

Figure 8.1 shows mask synthesis for DSAL. Contacts that are physically close in a layout are clustered (Fig. 8.1a) if their distance is smaller than a DSA distance rule, depending on the natural pitch of BCP [3–5]; ideal GP image that is expected to produce the desired contacts on a wafer after DSA process is determined through an inverse DSA [2, 6, 7] (Fig. 8.1b); a mask image is then synthesized through conventional optical proximity correction (OPC), which receives the ideal GP images as targets (Fig. 8.1c).

S. Shim and Y. Shin, *Physical Design and Mask Synthesis for Directed Self-Assembly Lithography*, NanoScience and Technology,
https://doi.org/10.1007/978-3-319-76294-4_8

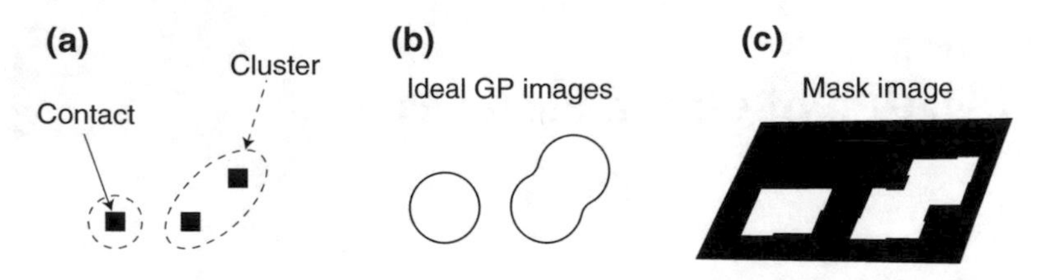

Fig. 8.1 DSAL mask synthesis: **a** contact clustering, **b** ideal GP images, and **c** a mask image

Even if the ideal GP image and its mask image are correctly synthesized, the GP that is created on a wafer may have some errors due to lithography variations, such as errors in scanner focus, exposure energy, and mask size [8]. Small errors of the GP shape may significantly affect final DSA pattern, and can even cause a patterning failure [9, 10]. It is very important to predict a problematic GP that will cause a pattern failure before manufacturing actual photomask, but it has relied on very lengthy DSA simulations so far [6, 11–13]. Based on quantitative experiments, for 10,000 GPs on a 100 μm × 100 μm contact layout, the verification based on the rigorous DSA simulation is estimated to take about 2 days (even with 100 multiple cores), which is not practical for a full-chip.

A verification function [1, 14] has been proposed to replace lengthy simulations and shorten the time for verification. It receives only a few geometric parameters of GP as variables, such as edge placement error (EPE), circles that help define GP contour, and sidewall angle; each GP in a given layout is represented as the parameters, which is then inputted to the verification function, and its output determines whether the GP can faithfully produce the desired contact holes with sufficient accuracy. The function is constructed as follows: a number of test GPs are prepared and each GP is tagged good or bad by assessment with DSA simulations in advance (see Fig. 8.2a); each GP is represented a few geometric parameters and mapped to a

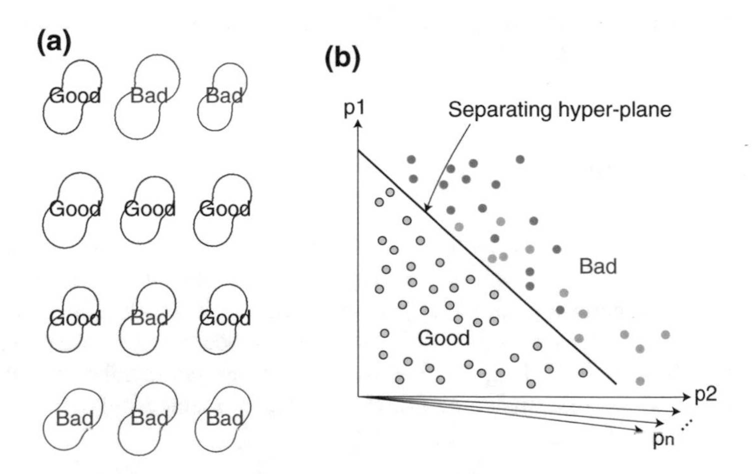

Fig. 8.2 Constructing the verification function: **a** DSA simulations of test GPs and **b** constructing a hyperplane that separates good and bad GPs

point in an n-dimensional space, where n is the number of the geometric parameters, and a hyperplane that separates good points and bad points is determined through a machine learning process. The resulting hyperplane, in turn, becomes a verification function (see Fig. 8.2b).

Key components in this process are summarized as follows:

- A systematic method of generating test GPs with comprehensive coverage of real GPs in a real layout.
- A small number of geometric parameters to characterize GP shape, and extracting a more compact parameter set through principal component analysis (PCA).
- Constructing a verification function: a set of functions, each of which is customized to a specific GP type, or a single global function, which can support any type of GP in a unified space.

The remainder of this chapter is organized as follows. Section 8.2 describes the test GP generation. Section 8.3 introduces a set of geometric parameters that are used to describe GPs and presents the use of PCA to reduce the size of that set. Section 8.4 gives a construction of the verification functions. In Sect. 8.5, application of this new method of GP verification is demonstrated using 10 nm technology. Section 8.6 draws conclusions.

8.2 Test GPs

8.2.1 *Preparation of GPs*

Assuming a gridded design rule (GDR) [15], all possible contact topologies can be enumerated, as shown in Fig. 8.3a. After instantiating each contact topology as a contact cluster, some neighboring contacts are randomly introduced so that the light interference that is likely to occur in a real design layout can be reproduced. The neighboring contacts are introduced in various configurations with different densities. Note that neighboring contacts are inserted a lithography pitch apart from the initial contact cluster being tested (see Fig. 8.3b), so that they are not clustered together. Contacts in the synthetic layout are clustered, and mask image is synthesized for the corresponding GPs [6, 7, 12]. This image is then inputted to lithography simulations with lithography variations, which output a set of litho images. Subsequent DSA simulations receive the litho images and yield DSA images of contacts, which can be compared with the required pattern of contacts, as shown in Fig. 8.3c. The GP whose error exceeding a tolerance is marked as bad (Fig. 8.3d).

In demonstrated tests, 10 nm technology with GDR [15] are assumed; the minimum contacted poly pitch (CPP) is 45 nm, the contact size is 22 nm by 22 nm, and the pitch of a metal track is 48 nm. ArF immersion (1.35 NA) lithography with annular illumination is assumed. The size of a synthetic layout clip is 2.5 μm $\times$ 2.5 μm, and a cluster under test is placed at the center of the clip; so, all neighboring GPs within an optical proximity range (about 1 μm) can be considered during lithography simulations. Mask image is synthesized by a modern OPC tool [16, 17], and the

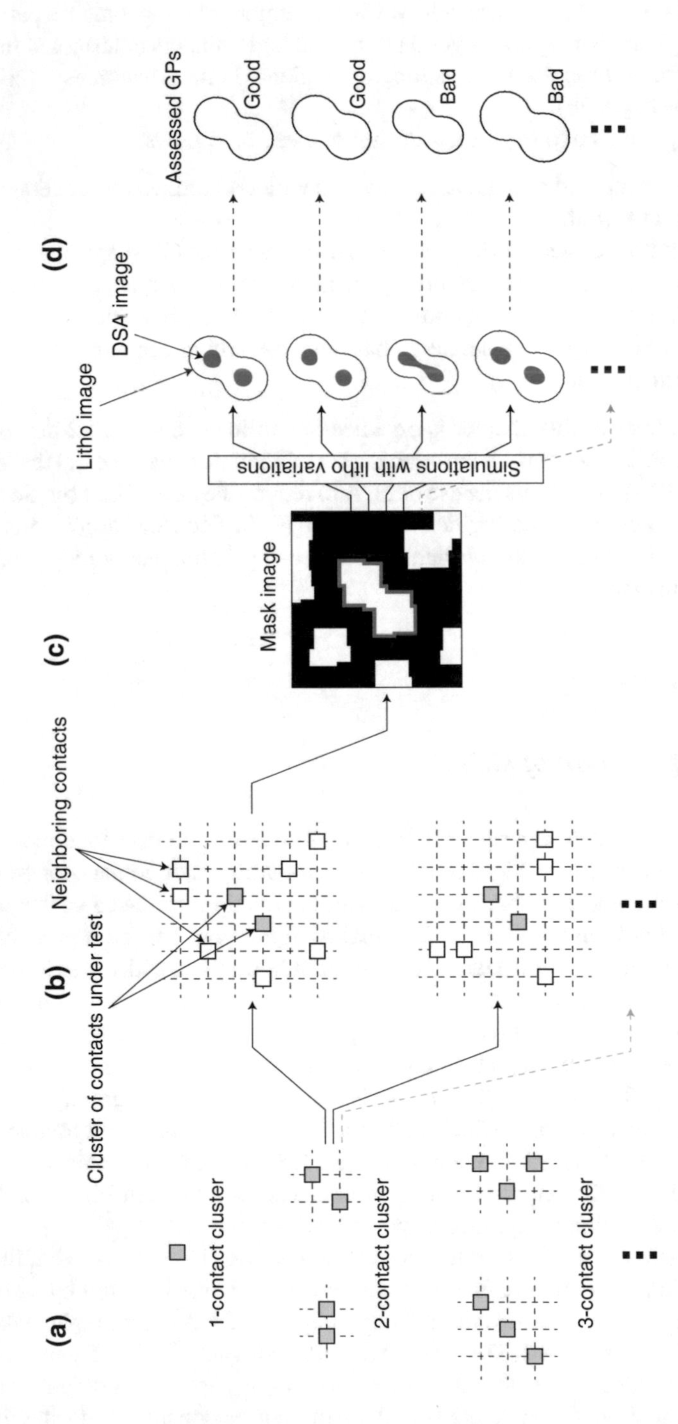

Fig. 8.3 Test GP preparations: **a** enumerating contact topologies, **b** introducing neighbor contacts, **c** litho- and DSA simulations after mask synthesis, and **d** assessing the GPs

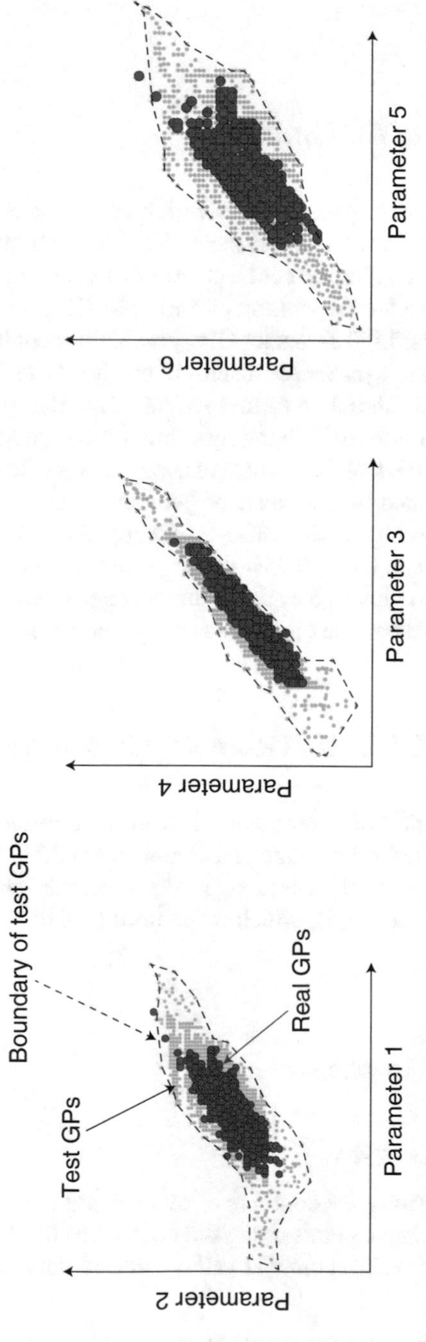

Fig. 8.4 Distributions of test and real GPs on three subspaces of n-dimensional parameter space

lithography simulations cover 27 extreme lithography conditions (combinations of two extreme and one nominal corners of scanner focus, exposure energy, and mask size [8, 18, 19]).

8.2.2 *Evaluation of GP Coverage*

Various contact clusters are generated, in which each cluster contains up to four contacts[1]; the total number of such cluster is 31. For each cluster of contacts, 10 different insertions of neighboring contacts with different densities are considered. This turns out to involve the preparation of 5400 test GPs, including 270 1-contact GPs, 540 2-contact GPs, 1350 3-contact GPs, and 3240 4-contact GPs.

Some test circuits are synthesized using 15 nm NanGate library [21], and the layouts are appropriately shrunk to follow virtual 10 nm design rules. Mask images for the test GPs are submitted to lithography simulations with lithography variations, which yield more than 100 k real GPs that are used to assess the coverage of test GPs.

Each GP is represented by a number of geometric parameters (details of these parameters will be presented in Sect. 8.3), and mapped to a point in n-dimensional space as shown in Fig. 8.4. Over 99.95% of the points corresponding to the real GPs reside within the region delineated by the points corresponding to the test GPs, which implies that the test GPs provide comprehensive coverage of the space of real GPs.

8.3 Preparing a GP Using Geometric Parameters

Three types of geometric parameter are used to abstract the shape of a GP: the plan view of a GP is characterized by edge placement error (EPE) as shown in Fig. 8.5a and inscribed and tangent circles as shown in Fig. 8.5b; the vertical shape of a GP is characterized by a sidewall angle, which is the third type of parameter, as shown in Fig. 8.5c.

8.3.1 *Geometric Parameters*

Edge Placement Error (EPE)

EPE is the distance between the contours of test GP and corresponding ideal GP. It is measured at a few sample points along the contour of ideal GP, such as points at peak (white dots in Fig. 8.5a), points at valley (shaded dots), and some more extra

[1]GPs containing five or more contacts are not manufacturable, so they are not allowed in real layouts [5, 20].

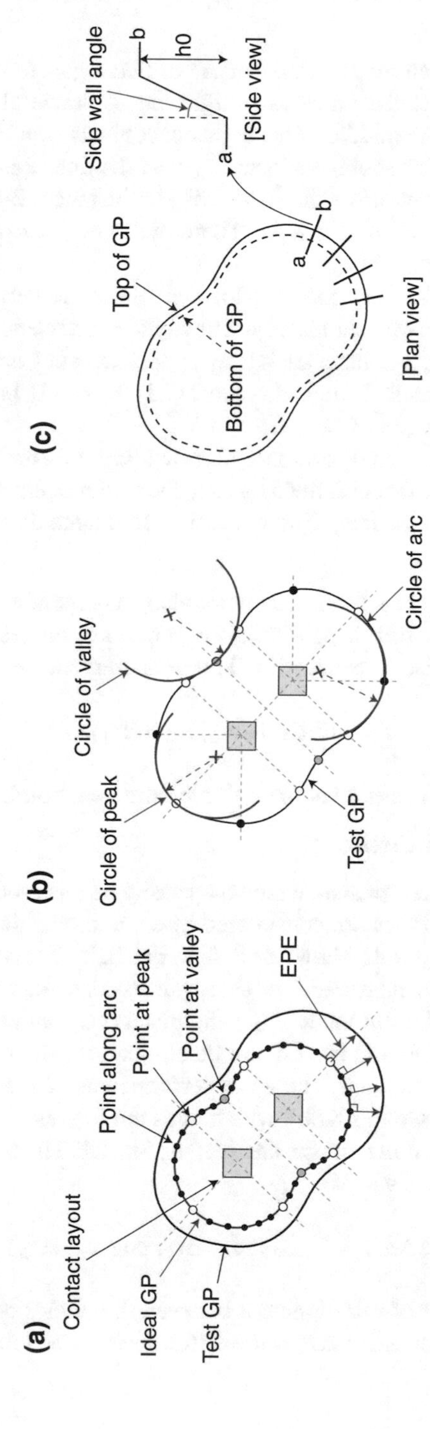

Fig. 8.5 Representation of a 2-contact GP using: **a** EPEs, **b** inscribed and tangent circles, and **c** side wall angles

points along the arcs between these points (black dots). Specifically, the points are determined as follows:

- A straight line is drawn through the centers of two adjacent contacts, and a line that is perpendicular to the previous straight line is drawn at the center of each contact as well as at the middle of the adjacent contacts (see Fig. 8.5a).
- The points at peak (white dots) are located where the lines that cross each contact center intersect the contour of the ideal GP. For instance, 2-contact GP has six points at peak as shown in Fig. 8.5a. There are $2(m + 1)$ points at peak if m contacts are lined up in a GP.
- The points at valley (shaded dots) are located where the perpendicular line that crosses the middle in between the adjacent contacts intersects the contour of the ideal GP. For instance, 2-contact GP has two points at valley as shown in Fig. 8.5a. A GP containing m linearly aligned contacts has $2(m - 1)$ points at valley, if m contacts are lined up in a GP.
- For remaining arcs of the GP contour, after dividing the arcs into segments with certain regular interval that is defined by user, the center of each segment is defined as point along arc (black dots). The number of segments determines the number of points along arc.

The EPEs are measured at each previous point along a perpendicular line to the ideal GP contour, such that an EPE is positive if the contour of the test GP is outside that of the ideal GP; otherwise, it is negative. The measurements are then arranged as a vector

$$\mathbf{v} = (\mathrm{EPE}_1, \mathrm{EPE}_2, ..., \mathrm{EPE}_n), \tag{8.1}$$

where EPE_k is an EPE measured at the k-th measurement point.

Inscribed and Tangent Circles

A 2-contact GP looks like a peanut, which may easily be represented by a set of arcs. Inscribed and tangent circles are constructed inside and outside the contour of test GP at the measurement points, shown as dots in Fig. 8.5b. For each circle, its center is identified and radius is measured. The same process is repeated for corresponding ideal GP. At every point, the amount of misalignment (i.e., center-to-center distance) of the circles for ideal GP and test GP are determined together with the difference between the radii between them. These are performed at 12 measurement points in case of a 2-contact GP (see Fig. 8.5b), which turns out to yield 24 parameters. There are $8(m+1)$ parameters if m contacts are lined up in a GP. The parameters measured for a GP are arranged as a vector

$$\mathbf{v} = (\Delta d_1, \Delta d_2, ..., \Delta d_n; \Delta r_1, \Delta r_2, ..., \Delta r_n), \tag{8.2}$$

where Δd_k is the extent of misalignment between the circles corresponding to the ideal and test GPs at k-th point; Δr_k is the difference between the radii of those two circles.

Sidewall Angle

Even if the contour of a GP is acceptable, it may cause undesirable contacts if the slope of GP wall is too small (or if GP sidewall angle is too large) [12, 22]. Two litho image contours are obtained at different image heights [23], one when contour forms on top of sidewall and the other when contour forms in the bottom. The sidewall angle (θ_k) at a measurement point k can be expressed by

$$\theta_k = \tan^{-1}\left(\frac{\mathrm{EPE}^{\mathrm{T}}{}_k - \mathrm{EPE}^{\mathrm{B}}{}_k}{h_0}\right), \quad (8.3)$$

where $\mathrm{EPE}^{\mathrm{T}}$ and $\mathrm{EPE}^{\mathrm{B}}$ are the edge placement error of the top and bottom contours, respectively, and h_0 is the height of sidewall. A number of sidewall angles of a single GP are arranged as a vector

$$\mathbf{v} = (\theta_1, \theta_2, ..., \theta_n). \quad (8.4)$$

Large sidewall angles usually occur when GPs are too close due to insufficient light contrast during optical lithography. The measurement points at which various parameters are measured are illustrated in Fig. 8.6. Table 8.1 gives the number of parameters for some GP types.

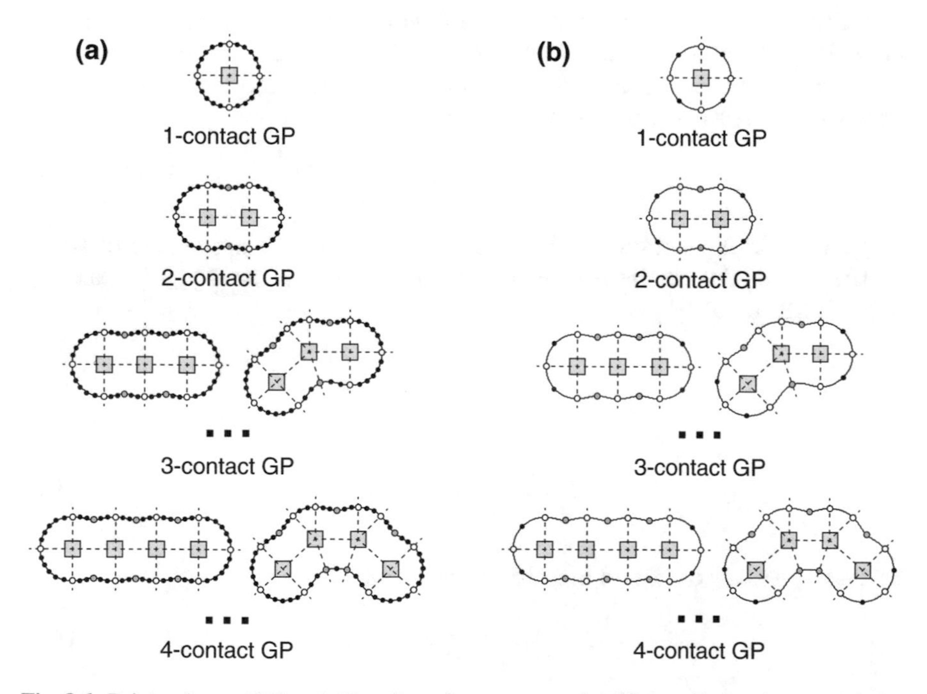

Fig. 8.6 Points where **a** EPE and sidewall angle are measured and **b** inscribed and tangent circles are constructed

Table 8.1 The number of parameters for each GP type and each set of parameters

GP type	Number of parameters	
	EPE & sidewall angle	Circles
1-contact GP	24	16
2-contact GP	36	24
3-contact GP	45–48	30–32
4-contact GP	48–60	28–40

8.3.2 *Principal Component Analysis*

A GP is now represented by a few geometric parameters. A principal component analysis (PCA) [24] may be applied to reduce the number of parameters so that constructing a verification function becomes easier. Specifically, PCA is applied to find a smaller set of new parameters, in a way that the variance of the parameters for GPs is maximized along the orthogonal axes in parameter space.

Suppose that there are m GPs, and the i-th GP is represented by a vector $\mathbf{x}_i \in \mathbb{R}^n$, where n is the number of original geometric parameters. A set of m vectors will constitute a matrix $\mathbf{X} \in \mathbb{R}^{m\times n}$, in which each vector corresponds to a row. Mean of each column is calculated, and each element in that column is subtracted by the mean, i.e., column contains the difference from the mean. Let $\mathbf{y}_i$ be a vector in a new space ($\mathbb{R}^n$) corresponding to $\mathbf{x}_i$. The k-th element of $\mathbf{y}_i$ is obtained by projecting $\mathbf{x}_i$ on to the k-th axis in the new space as follows:

$$y_{i,k} = \mathbf{x}_i \mathbf{w}_k, \tag{8.5}$$

where $\mathbf{w}_k \in \mathbb{R}^n$ is a unit vector which corresponds to the k-th axis in the new space.

The axis $\mathbf{w}_1$ that maximizes the variance of $y_{1,1}$, $y_{2,1}$, ..., and $y_{m,1}$ is searched. Such $\mathbf{w}_1$ has to satisfy

$$\begin{aligned} \mathbf{w}_1 &= \arg\max_{\mathbf{w}} \sum_{i=1}^{m} y_{i,1}^2 \\ &= \arg\max_{\mathbf{w}} \sum_{i=1}^{m} (\mathbf{x}_i \mathbf{w})^2, \end{aligned} \tag{8.6}$$

which may also be expressed in matrix form, which is given by

$$\mathbf{w}_1 = \arg\max_{\mathbf{w}} (\mathbf{w}^{\mathrm{T}} \mathbf{X}^{\mathrm{T}} \mathbf{X} \mathbf{w}). \tag{8.7}$$

Solving (8.7) is equivalent to eigenvalue decomposition of a symmetric matrix $\mathbf{X}^{\mathrm{T}}\mathbf{X} \in \mathbb{R}^{n\times n}$, which yields the eigenvalue $\mathbf{w}^{\mathrm{T}}\mathbf{X}^{\mathrm{T}}\mathbf{X}\mathbf{w}$ when $\mathbf{w}$ is the corresponding eigenvector. In particular, $\mathbf{w}_1$ is the eigenvector associated with the largest eigenvalue (i.e., the largest variance). The eigenvectors are then sorted by their eigenvalues in descending order, and the first l eigenvectors (i.e., $\mathbf{w}_1$, $\mathbf{w}_2$, ..., and $\mathbf{w}_l$) are selected as new parameters; the selected eigenvectors are called principal components (PCs). Once a new set of l PCs is determined, the ith GP is now represented by a vector $\mathbf{y}_i \in \mathbb{R}^l$ as follows:

$$\mathbf{y}_i^{\mathrm{T}} = \begin{bmatrix} \mathbf{x}_i \mathbf{w}_1 \\ \mathbf{x}_i \mathbf{w}_2 \\ \cdots \\ \mathbf{x}_i \mathbf{w}_l \end{bmatrix}^{\mathrm{T}} = \mathbf{x}_i^{\mathrm{T}} \mathbf{W}, \quad (8.8)$$

where $\mathbf{W}$ is n-by-l matrix, whose ith column corresponds to $\mathbf{w}_i$.

8.3.3 Experimental Observations

The test GPs mentioned in Sect. 8.2.2 are represented by the three types of parameters, and they are mapped to the points in n-dimensional parameter space. The variance of this point-set along each axis is calculated, and PCA is applied to them.

For example, a 2-contact GP is represented by 36 EPEs, which are mapped to a point in 36-dimensional parameter space as shown in Fig. 8.7a. Such point-set is then projected into the space of a few PCs as shown in Fig. 8.7b. The EPEs and PCs are sorted by their variance, which are compared as shown in Fig. 8.8a. No EPEs are outstanding in variance value, meaning that all EPEs should be kept. A PC is, in essence, a linear combination of all EPEs. Therefore, depending on weighting

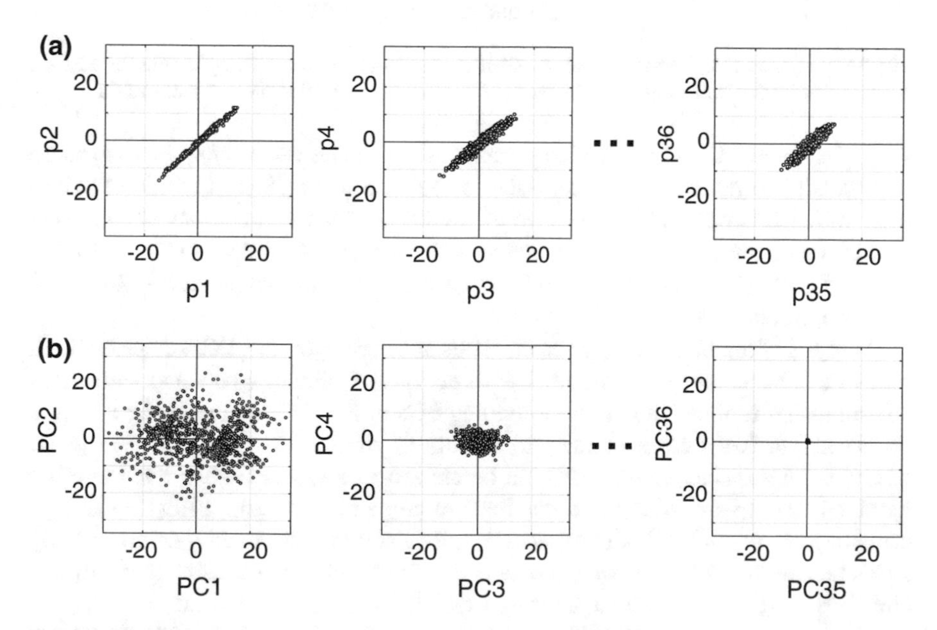

Fig. 8.7 Distribution of test GPs represented by **a** EPEs and **b** PCs

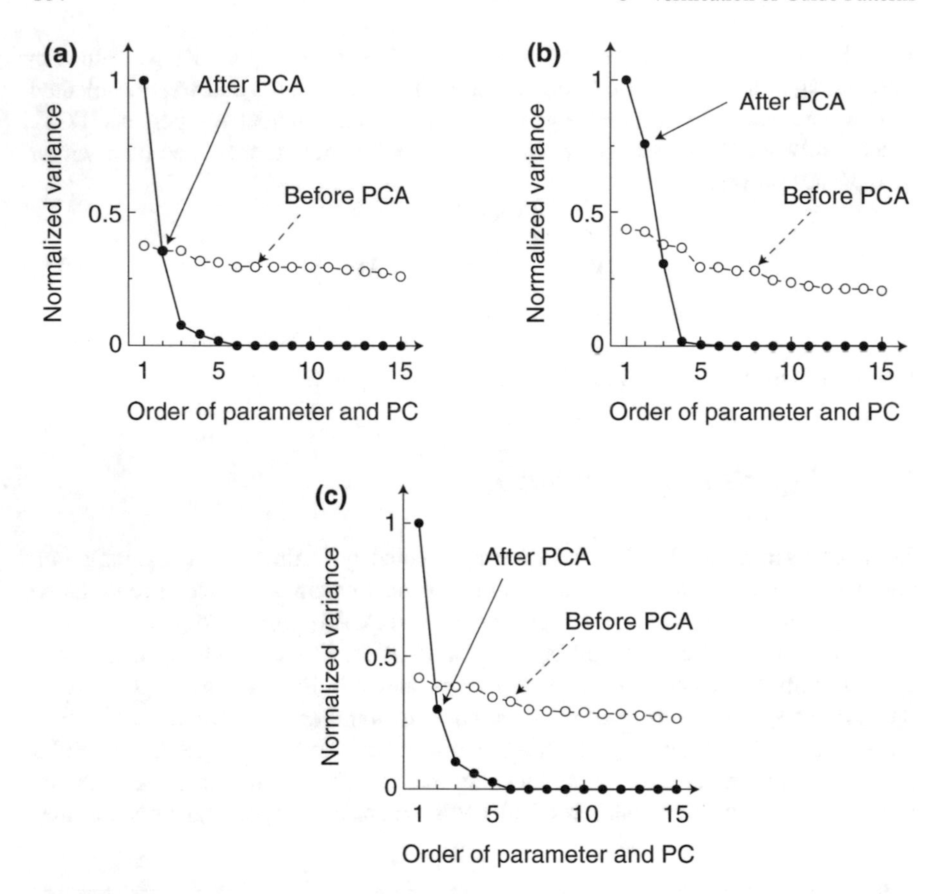

Fig. 8.8 Variance of 15 parameters before and after PCA, for 2-contact GPs represented by **a** EPEs, **b** circles, and **c** side wall angles. All variance values are normalized to the variance of PC1

of EPEs, some PCs may have large variance and some others may be associated with small insignificant variance. Such is the case in Fig. 8.8a, in which 98% of total variance is occupied by top five PCs with largest variance; they can be used to construct a small set of PC parameters. For parameters of circles (Fig. 8.8b) and sidewall angle (Fig. 8.8c), three and five PCs are sufficient to represent 98% of total variance, respectively.

Analysis of relationship between the EPEs and corresponding PCs suggests which part of a GP contour is important. The directions of vectors associated with EPEs are compared to those of the corresponding PCs in PC_1-PC_2 space in Fig. 8.9. It is observed that EPEs corresponding to vectors of similar direction, e.g., $EPE_{17\sim21}$, are highly correlated, so one EPE can be chosen as a representative. On the other hand, EPEs corresponding to vectors that are perpendicular, e.g., EPE_1 and EPE_{17}, are independent, and so both are necessary. PC_1 and PC_2 are mainly determined by EPEs that are measured at two extreme ends of GP contour, respectively ($EPE_{17\sim21}$ and $EPE_{1\sim3,35,36}$); this fact suggests that the EPEs measured at points 1, 7, 10, 13, and 19 are most important and should be carefully considered during GP patterning.

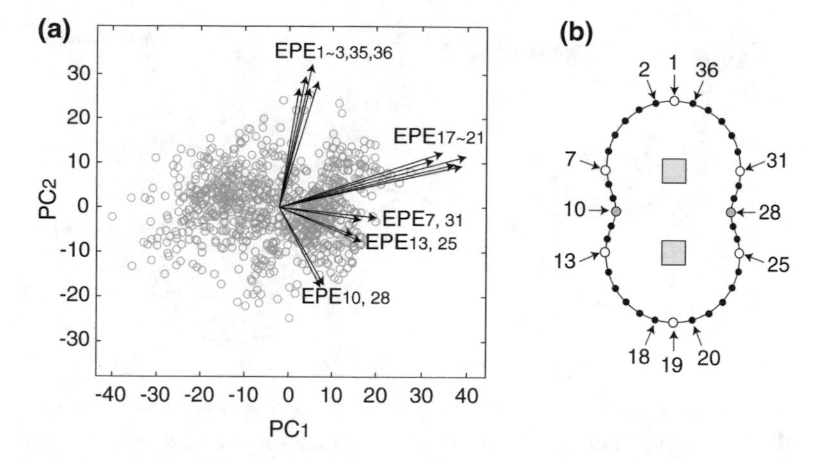

Fig. 8.9 **a** Vectors associated with EPEs in space of PC_1-PC_2 and **b** the points where these EPE values are measured

The other parameter types can also be analyzed in similar fashion. It is observed that two tangent circles at point 10 and 28 and two inscribed circles at point 1 and 19 are most important, which also indicates that those points are important for DSA process. Measurement of sidewall angle is most significant at such points that are considered to be important for EPEs.

8.4 Constructing a Verification Function

Each test GP can be mapped to a point in n-dimensional space, and be tagged "good" or "bad". A hyperplane that separates the good and bad points can be used as a verification function. A support vector machine (SVM) is a technique that can be used to determine such hyperplane while the margin that separates the good points from the bad points is maximized, as shown in Fig. 8.10a [25].

It turns out that the good points cannot be separated from the bad points by a typical linear function with any accuracy because the parameter values are radially distributed; the good points are usually distributed near the origin of the space, and the bad points surround them, as shown in Fig. 8.10b. A kernel trick can be applied, in which a kernel function such as polynomial function and Gaussian radial basis function (RBF) is used to map the original space to another. Assume that Gaussian RBF is used, which is more flexible. Gaussian RBF is given by

$$k(\mathbf{x}_i, \mathbf{x}_j) = \exp\left(-\frac{||\mathbf{x}_i - \mathbf{x}_j||^2}{2\sigma^2}\right), \tag{8.9}$$

where $\mathbf{x}_i$ is a vector corresponding to ith point and σ is a parameter that determines the kernel width. A point $\mathbf{x}_i$ in the original space is mapped to the one in higher dimensional space through the following function:

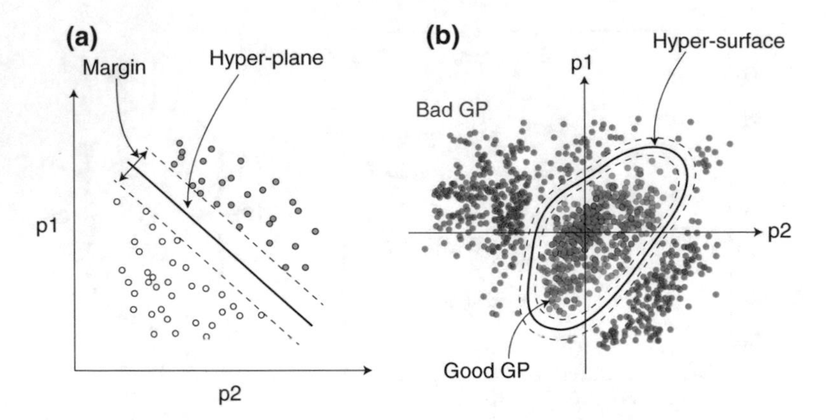

Fig. 8.10 Constructing a verification function in two-dimensional space: **a** a typical hyperplane in a linear space and **b** a hypersurface in a nonlinear space that occurs in GP verification

$$f(\mathbf{x}_i) = \sum_{j=1}^{N} y_j k(\mathbf{x}_i, \mathbf{x}_j), \tag{8.10}$$

where N is the size of point-set; y_j is 1 if $\mathbf{x}_j$ is a bad point and it is -1 otherwise. Figure 8.11 shows an example of data distribution before and after applying kernel trick. Points in original two-dimensional space are mapped to those in three-dimensional space, and a hyperplane in the three-dimensional space takes a role of a hypersurface in two-dimensional space.

A hyperplane is determined by a quadratic programming formulation of SVM, whose dual form is given by [25, 26]

$$\text{Maximize } \sum_{i}^{N} \alpha_i - \frac{1}{2} \sum_{i}^{N} \sum_{j}^{N} \alpha_i \alpha_j y_i y_j k(\mathbf{x}_i, \mathbf{x}_j), \tag{8.11}$$

which is subject to

$$\sum_{i}^{N} y_i \alpha_i = 0, \tag{8.12}$$

$$0 \geq \alpha_i \geq C, \tag{8.13}$$

where α_i and C, respectively, are a Lagrange multiplier and its value limit in a certain boundary; $k(\mathbf{x}_i, \mathbf{x}_j)$ indicates the Gaussian RBF (8.9), which leads to an optimal classification it is symmetric positive semidefinite.

The width of kernel function, which depends on σ and C, determines how specific the verification function is to the test GPs; for instance, too narrow a function may cause overfitting. They are adjusted by tenfold cross-validation. A set of test GPs is randomly split into 10 subsets of equal size: nine of them are used to construct

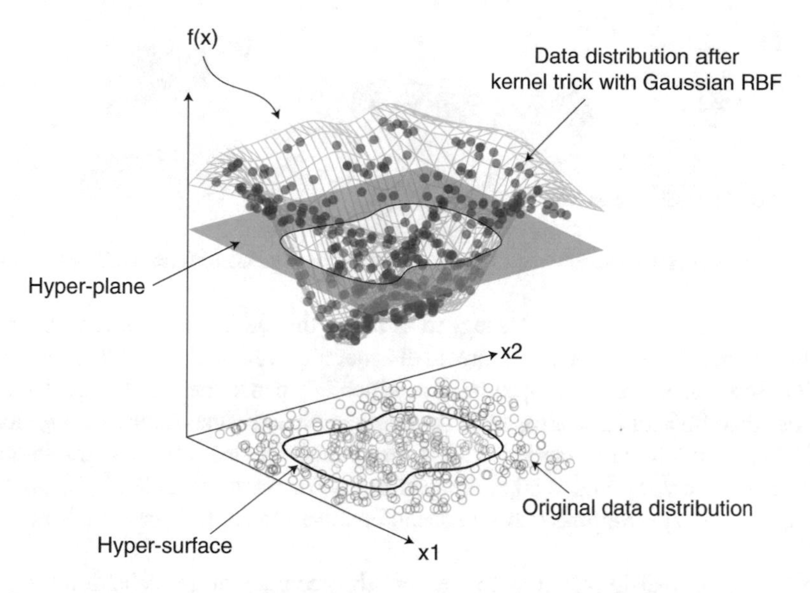

Fig. 8.11 Original data in two-dimensional space is mapped to data in three-dimensional space after applying kernel trick with a Gaussian RBF. A hyperplane plays a role of a hypersurface in the process

a verification function and a remaining set is used to assess the accuracy of the function. This process is repeated 10 times with different combinations of subset, and the accuracy is averaged over the 10 iterations. This adjustment procedure is embedded into an optimization process that determines the value of σ and C with the highest accuracy. Note that different GP types cannot be handled in the same parameter space due to different numbers of parameters, so a verification function is constructed for each GP type.

In actual verification stage, GP type is first identified by checking the contact layout topology of an unknown GP; the unknown GP is then parameterized in the same manner of Sect. 8.3. A vector made of the parameters is submitted to the verification function of corresponding GP type, which determines whether the GP is good or bad.

8.5 Experimental Assessment

Representation of GP using geometric parameters, application of PCA, and construction of a verification function are implemented using MATLAB. Verification functions are constructed using 1350 test GPs, which consist of three GP types including 270 1-contact GPs, 540 2-contact GPs, and 540 (linearly aligned) 3-contact GPs, as shown in Fig. 8.12a. For verification purpose, another 2000 unknown GPs are prepared, in which 1350 are generated as described in Sect. 8.2, and 650 are extracted from a few real contact layouts. These unknown GPs include new GP types, e.g., GPs with three contacts that are located in zigzag fashion and GPs with four contacts, as shown in Fig. 8.12b. A DSA simulator based on self-consistent field theory

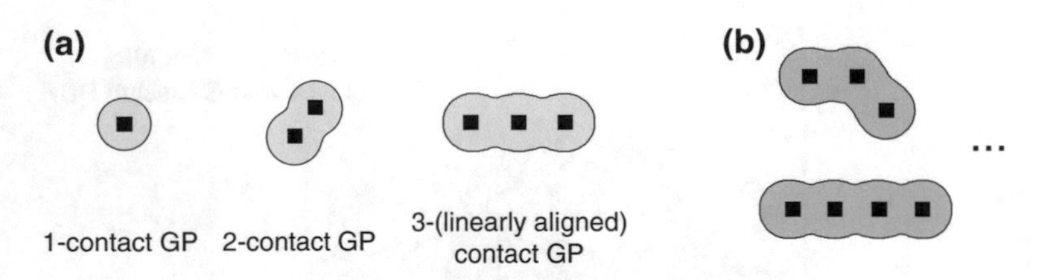

Fig. 8.12 **a** Three GP types and **b** new GP types with different geometries from (**a**) for verification

(SCFT) [13, 27] is used. A GP is tagged as bad if the DSA simulator detects bridge failure, which is when one or more of its member contacts are 10% larger than their desired size, or contact open failure, which is when some member contacts are smaller than 80% of its desired area; the exact criteria of such failures are typically available from a foundry fab [28, 29]. A simulation for one test GP takes about 15 minutes on average,[2] and it takes about 8 hours for whole test GPs if 50 multiple cores are used; this runtime may be acceptable given that whole process is required just once.

Recall and precision are used to measure the accuracy of a verification function. Recall indicates how many bad GPs are correctly detected. It is given by

$$\text{Recall} = \frac{\#\text{True positives}}{\#\text{Bad GPs}}, \tag{8.14}$$

where the detection of a bad GP is named a true positive (see Fig. 8.13). Precision measures the confidence of detection of bad GPs. It is defined by

$$\text{Precision} = \frac{\#\text{True positives}}{\#\text{True positives} + \#\text{False negatives}}, \tag{8.15}$$

where false negative indicates a good GP that is incorrectly verified as bad GPs. In practice, it is more important to improve recall for the safe manufacturing, which is however sometimes associated with low precision. Specifically, if too many good GPs are detected as bad GPs (i.e., false negatives), it causes unnecessary effort of mask resynthesis and design refinement to avoid such false negatives.

8.5.1 Choice of Parameters

A verification function is constructed for the test GPs of 1, 2, and 3 (linearly aligned) contacts, respectively. For each GP type, various combinations of parameters are tried to understand the impact of each parameter on the accuracy of verification function. Table 8.2 shows the accuracy of recall and precision values while unknown GPs are verified by the functions.

[2] A DSA simulation for 1-, 2-, and 3-contact GP takes about 5, 7, and 20 minutes, respectively.

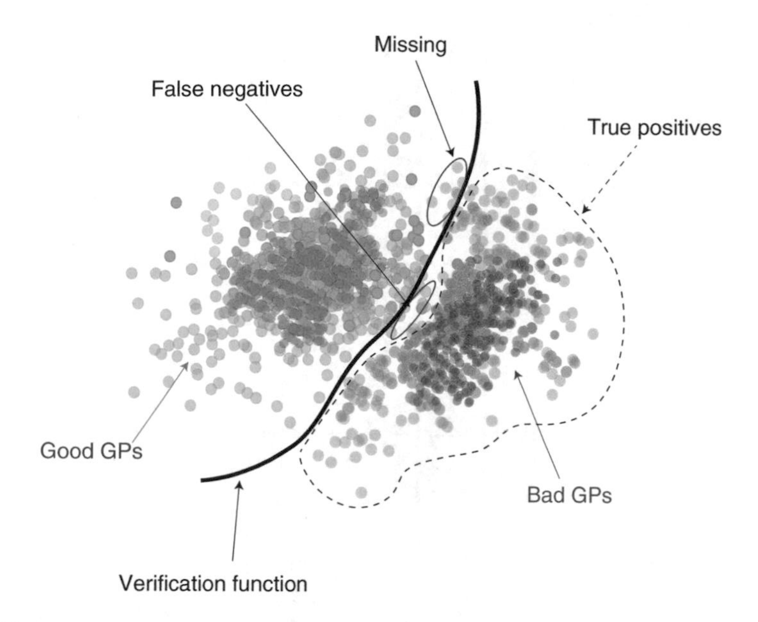

Fig. 8.13 Result of GP verification. Some good GPs are incorrectly detected (false negative), and some bad GPs are not detected (missing)

Contribution of sidewall angle to accuracy is significant independently from that of other parameters, because the sidewall angle is geometrically orthogonal to others. If sidewall angle is not taken as a parameter (see second and fourth rows corresponding to EPE and Circle), recall is very low, implying many bad GPs are undetected. Considering sidewall angle together with EPE or circle (see third and fifth rows) improves recall. Taking all parameters into account (see the last row) achieves the highest recall (97.9%) and precision (95.2%) over the whole set of GPs.

The numbers in parentheses indicate the accuracy of verification function when only unknown GPs extracted from real layouts are verified. This is not that different from the overall accuracy when whole unknown GPs are applied, which implies that the test GPs used for function construction cover the unknown GPs very well, as demonstrated in Sect. 8.2.2.

8.5.2 Parameter Reduction

Application of PCA is assessed in its impact on a verification function. The same set of GPs from Sect. 8.5.1 is taken, all geometric parameters (EPE, circles, and sidewall angle) are used to represent each GP, and PCA is applied. A set of PCs that represent 98% of total variance is identified. The number of such PCs is shown in the third column of Table 8.3 for each GP type; the number of original geometric parameters is also shown and compared.

Table 8.2 Accuracy of verification functions with various combinations of parameters (all numbers are percentage). The numbers in parentheses indicate the accuracy for only unknown GPs extracted from real layouts

Parameter sets	1-contact GP		2-contact GP		3-contact GP		All GPs	
	Recall	Precision	Recall	Precision	Recall	Precision	Recall	Precision
EPE	85.4 (83.9)	90.7 (88.4)	89.0 (87.1)	85.0 (85.1)	89.9 (88.7)	92.4 (90.9)	89.0 (87.8)	90.1 (89.0)
EPE + angle	95.1 (95.8)	91.6 (89.7)	95.9 (96.3)	88.7 (88.2)	97.1 (96.7)	94.3 (94.7)	96.5 (95.9)	91.9 (90.2)
Circle	82.5 (81.2)	88.5 (86.1)	84.3 (82.9)	82.4 (81.8)	89.1 (90.4)	91.3 (91.9)	86.8 (87.2)	88.4 (87.7)
Circle + angle	96.1 (95.9)	88.4 (88.6)	95.3 (94.8)	86.6 (86.0)	97.6 (97.0)	93.6 (94.0)	96.8 (96.9)	90.8 (90.1)
EPE+circle+angle	95.1 (94.6)	94.2 (93.1)	97.1 (96.9)	93.8 (93.0)	98.4 (97.9)	96.1 (95.8)	97.9 (98.0)	95.2 (94.6)

Table 8.3 Comparison of verification functions of geometric parameters and PCs

GP type	Parameter set	#Parameters	Accuracy (%)		Runtime (sec)	
			Recall	Precision	Construction	Verification
1-contact GP	Whole	64	95.1	94.2	37	0.1
	PCs	12	97.1	96.2	11	0.1
2-contact GP	Whole	96	97.1	93.8	289	0.1
	PCs	13	97.7	96.0	38	0.1
3-contact GP	Whole	128	98.4	96.1	844	0.1
	PCs	13	98.9	97.1	33	0.1
All	Whole	–	97.9	95.2	1170	0.1
	PCs	–	98.3	96.7	82	0.1

Table 8.4 Comparison of four GP verification methods (all numbers are percentage)

GP type	Verification function		AdaBoost		SVR		CDSA model	
	Recall	Precision	Recall	Precision	Recall	Precision	Recall	Precision
1-contact GP	97.1	96.2	92.5	92.1	96.4	97.8	98.7	97.1
2-contact GP	97.7	96.0	90.2	86.5	92.7	91.9	94.1	91.3
3-contact GP	98.9	97.1	91.6	88.7	91.7	90.1	92.6	90.3
All	98.3	96.7	90.8	88.1	92.5	91.4	93.3	91.2

The reduced number of parameters through PCA affects the runtime to construct a verification function. As shown in column 6, runtime is reduced to 7%, on average of all GP types, of time required to construct a function with all geometric parameters. Actual GP verification, which is quick, is not noticeably affected by the number of parameters as shown in the last column.

Using only a few PCs helps accuracy, both in recall and precision, rather than sacrifices it. This is because PCs highlight the difference of GPs and so the verification function can tell the good GPs from the bad GPs more easily.

8.5.3 Comparison of GP Verification Methods

GP verification using verification function is compared to three other methods. In AdaBoost method [10], three parameters (point correspondence, histogram of oriented gradients, and segment distance) are used to represent a GP. A verification model is constructed for each GP type, respectively. Recall and precision shown in fourth and fifth column of Table 8.4) are lower than those of verification function (second and third column) by 7.5 and 8.6%, respectively. This method uses significantly larger number of parameters, e.g., 1761 parameters for a 2-contact GP as opposed to only 13 in the method of verification function. As a result, model construction takes 17× more time.

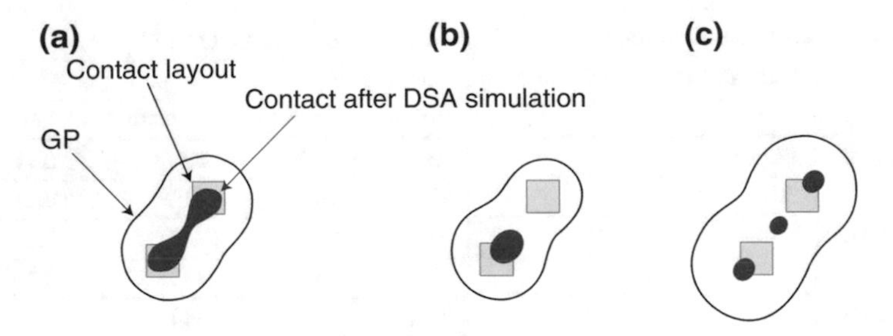

Fig. 8.14 Example of DSA phase transition: **a** contact bridge, **b** contact not open, and **c** unexpected pattern

A support vector regression (SVR) model [30] and compact DSA (CDSA) model [9, 31–34] have been proposed to predict the pitch and location of final DSA contacts. Both can also be compared to the method of verification function as well as to AdaBoost method. In SVR model, the same parameters used for AdaBoost are employed again. The models are calibrated for each type of test GPs; note that because GP with large error causing DSA phase transition (Fig. 8.14) cannot be handled, such GPs are dropped from the calibration process. The calibrated models are applied to unknown GPs and predict their pitch and location of contacts well unless GP error is large to cause pattern failure; however, prediction accuracy significantly decreases as GP error increases. As shown in Table 8.4, SVR-based model yields recall and precision that is lower than those of the method of verification function by 5.8 and 5.3%, respectively. The accuracy of CDSA model is also inferior, by 5 and 5.5%, respectively.

8.5.4 A Global Verification Function

Instead of using some verification functions customized to each particular GP types, a single global verification function may be introduced to accommodate all GP types. Now, all GPs should be represented in a unified fashion in the same parameter space:

$$\mathbf{v_i} = (\mathrm{EPE}_p, \mathrm{EPE}_v, \mathrm{EPE}_a; \Delta r_p, \Delta r_v, \Delta r_a, \Delta d_p, \Delta d_v, \Delta d_a; \theta_p, \theta_v, \theta_a), \tag{8.16}$$

where EPE_p, Δr_p, Δd_p, and θ_p correspond to the maximum EPE, maximum difference of radius of circles, maximum misalignment of circles, and maximum sidewall angle that is only measured at points of peak, respectively; subscripts v and a indicate valleys and arcs, respectively. Note that because 1-contact GP has no valley, all corresponding entries are set to zero. It can be shown that any type of GP can be represented by using 12 parameters in (8.16), so a single global function indeed exists.

A global verification function is constructed in PC space using the same test GPs used in Sect. 8.5.2. The function is applied to the whole unknown GPs, in

Table 8.5 Accuracy of custom and global verification functions for basic GP type and new GP type

	Accuracy	Basic GP type	New GP type
Custom functions	Recall (%)	98.3	–
	Precision (%)	96.7	–
Global function	Recall (%)	97.9	97.6
	Precision (%)	87.1	83.9

which 60% belong to the basic types (Fig. 8.12a) and 40% are GPs from new contact topologies (Fig. 8.12b). The accuracy of global verification function is compared to that of custom functions in Table 8.5. The recall and precision values of custom functions are simply taken from Table 8.3 (all GP type and when PCs are used for parameters). Custom and global verification functions have not much different recall (98.3% versus 97.9%), but global verification function has 9% lower precision than custom function. This is because a global function is somewhat pessimistic, since only maximum parameter values are considered to represent a GP in an effort to unify the parameter space. When the global function is applied to new GP types, recall is not much different but precision becomes only 83.9%. Custom verification functions cannot be applied to new GP types, so no comparison is made.

Global and custom verification functions may be used for different purposes. Global function might be useful in the early stage of layout and process development, when many unexpected GP types may be encountered. In this stage, applicability of function is more important than absolute accuracy because various test layouts are usually tried to develop process and design rules. In later stage of development, e.g., during mass production, design rules are already fixed, which allows only a few contact topologies for DSAL compliant layout [5, 20]. As a result, all GP types can be enumerated, and custom verification function can be constructed for each GP type to provide better accuracy.

8.6 Conclusions

A GP cannot be analytically synthesized for a given group of contacts, so its construction is usually based on empirical approach. Therefore, a verification of GPs to ensure all contacts are correctly patterned is very important. It has been shown that a verification function can be constructed and be used instead of lengthy simulations. The function is constructed in following three steps: preparing necessary yet a small number of test GPs; representing each GP using a small number of geometric parameters, which can be even reduced by applying PCA; constructing the verification function through a support vector machine. It has been demonstrated that this approach produces acceptable verification accuracy in 10 nm technology.

References

1. S. Shim, S. Cai, J. Yang, S. Yang, B. Choi, Y. Shin, Verification of directed self-assembly (DSA) guide patterns through machine learning, in *Proceedings for SPIE Advanced Lithography* (2015), pp. 1–8
2. L. Azat, Computational solution of inverse directed self-assembly problem, in *Proceedings for SPIE Advanced Lithography* (2013), pp. 1–10
3. M. Muramatsu, M. Iwashita, T. Kitano, T. Toshima, M. Somervell, Y. Seino, D. Kawamura, M. Kanno, K. Kobayashi, T. Azuma, Nanopatterning of diblock copolymer directed self-assembly lithography with wet development. J. Micro Nanolithogr. MEMS MOEMS **11**(3), 1–6 (2012)
4. Y. Seino, H. Yonemitsu, H. Sato, M. Kanno, H. Kato, K. Kobayashi, A. Kawanishi, T. Azuma, M. Muramatsu, S. Nagahara, T. Kitano, T. Toshima, Contact hole shrink process using graphoepitaxial directed self-assembly lithography. J. Micro Nanolithogr. MEMS MOEMS **12**(3), 1–6 (2013)
5. H. Yi, X. Bao, R. Tiberio, P. Wong, Design strategy of small topographical guiding templates for sub-15nm integrated circuits contact hole patterns using block copolymer directed self-assembly, in *Proceedings for SPIE Advanced Lithography* (2013), pp. 1–9
6. W. Wang, L. Azat, Y. Zou, T. Coskun, A full-chip DSA correction framework, in *Proceedings for SPIE Advanced Lithography* (2014), pp. 1–11
7. S. Shim, Y. Shin, Mask optimization for directed self-assembly lithography: inverse DSA and inverse lithography, in *Proceedings of the Asia South Pacific Design Automation Conference* (2016), pp. 83–88
8. L. Liebmann, S. Mansfield, G. Han, J. Culp, J. Hibbeler, R. Tsai, Reducing DfM to practice: the lithography manufacturability assessor, in *Proceedings for SPIE Advanced Lithography* (2006), pp. 786–798
9. G. Fenger, A. Burbine, J. Torres, Y. Ma, Y. Granik, P. Krasnova, G. Vandenberghe, R. Gronheid, J. Bekaert, Calibration and application of a DSA compact model for grapho-epitaxy hole processing using contour-based metrology, in *Proceedings for SPIE Advanced Lithography* (2014), pp. 1–12
10. Z. Xiao, Y. Du, H. Tian, M. Wong, H. Yi, H. Wong, H. Zhang, Directed self-assembly (DSA) template pattern verification, in *Proceedings of the Asia South Pacific Design Automation Conference* (2014), pp. 1–6
11. H. Yi, L. Azat, P. Wong, Computational simulation of block copolymer directed self-assembly in small topographical guiding templates, in *Proceedings for SPIE Advanced Lithography* (2013), pp. 1–7
12. L. Azat, G. Grant, P. Moshe, S. Gerard, W. Wong, J. Xu, Y. Zou, Computational simulations and parametric studies for directed self-assembly process development and solution of the inverse directed self-assembly problem. Jpn. J. Appl. Phys. **53**(6S), 1–8 (2014)
13. N. Laachi, K.T. Delaney, B. Kim, S. Hur, R. Bristol, D. Shykind, C.J. Weinheimer, G.H. Fredrickson, Self-consistent field theory investigation of directed self-assembly in cylindrical confinement. J. Polym. Sci. Part B Polym. Phys. **53**(2), 142–153 (2015)
14. S. Shim, Y. Shin, Fast verification of guide patterns for directed self-assembly lithography, IEEE Trans. CAD Integr. Circuits Syst. to be published
15. M.C. Smayling, V. Axelrad, 32nm and below logic patterning using optimized illumination and double patterning, in *Proceedings for SPIE Advanced Lithography* (2009), pp. 1–10
16. *Calibre nmOPC Manual*, Mentor Graphics (2013)
17. *Calibre OPCverify Manual*, Mentor Graphics (2013)
18. K. Peter, R. Marz, S. Grondahl, W. Maurer, Litho-friendly design (LfD) methodologies applied to library cells, in *Proceedings for SPIE Advanced Lithography* (2013), pp. 1–9
19. J.P. Cain, Design for manufacturability: a fabless perspective, in *Proceedings for SPIE Advanced Lithography* (2013), pp. 1–9
20. Y. Du, D. Guo, M. Wong, H. Yi, H. Wong, H. Zhang, Q. Ma, Block copolymer directed self-assembly (dsa) aware contact layer optimization for 10nm 1D standard cell library, in *Proceedings of International Conference on Computer Aided Design* (2013), pp. 186–193

21. Nangate 15nm open cell library, http://www.nangate.com/
22. K. Lai, C. Liu, J. Pitera, D.J. Dechene, A. Schepis, J. Abdallah, H. Tsai, M. Guillorn, J. Cheng, G. Doerk, M. Tjio, C. Rettner, O. Odesanya, M. Ozlem, N. Lafferty, Computational aspects of optical lithography extension by directed self-assembly, in *Proceedings for SPIE Advanced Lithography* (2013), pp. 1–10
23. Y. Tang, C. Chou, W. Huang, R. Liu, T. Gau, Multiple-image-depth modeling for hotspot and AF printing detections, in *Proceedings for SPIE Advanced Lithography* (2012), pp. 1–7
24. I.T. Jolliffe, *Principal Component Analysis*, (Springer, 2002)
25. Q. Wu, D.-X. Zhou, SVM soft margin classifiers: linear programming versus quadratic programming. Neural Comput. **17**(5), 1160–1187 (2005)
26. T. Hofmann, B. Schölkopf, A.J. Smola, Kernel methods in machine learning. Ann. Statist. **36**(3), 1171–1220 (2008)
27. H.D. Ceniceros, G.H. Fredrickson, Numerical solution of polymer self-consistent field theory. Multiscale Model. Simul. **2**(3), 452–474 (2004)
28. S. Paek, J. Kang, N. Ha, B. Kim, Yield enhancement with DFM, in *Proceedings for SPIE Advanced Lithography* (2012), pp. 128–138
29. Samsung Electronics Corp. OPC principal engineer, *Personal Communication* (2014)
30. Z. Xiao, Y. Du, M. Wong, H. Yi, H. Wong, H. Zhang, Contact pitch and location prediction for directed self-assembly template verification, in *Proceedings of the Asia South Pacific Design Automation Conference* (2015), pp. 644–651
31. J. Torres, S. Kyohei, D. Fryer, Y. Granik, Y. Ma, P. Krasnova, G. Fenger, S. Nagahara, S. Kawakami, B. Rathsack, G. Khaira, J. Pablo, J. Ryckaert, Physical verification and manufacturing of contact/via layers using grapho-epitaxy DSA process, in *Proceedings for SPIE Advanced Lithography* (2014), pp. 1–8
32. G. Fenger, J. Torres, Y. Ma, Y. Granik, P. Krasnova, A. Fouquet, J. Belledent, A. Gharbi, R. Tiron, Compact model experimental validation for grapho-epitaxy hole processes and its impact in mask making tolerances, in *Proceedings for SPIE Advanced Lithography* (2014), pp. 1–11
33. Y. Ma, J. Lei, J.A. Torres, L. Hong, J. Word, G. Fenger, A. Tritchkov, G. Lippincott, R. Gupta, N. Lafferty, Y. He, J. Bekaert, G. Vanderberghe, Directed self-assembly (DSA) grapho-epitaxy template generation with immersion lithography, in *Proceedings for SPIE Advanced Lithography* (2015), pp. 1–11
34. Y. Ma, Y. Wang, J. Word, J. Lei, J. Mitra, J.A. Torres, L. Hong, G. Fenger, D. Khaira, M. Preil, L. Yuan, J. Kye, H.J. Levinson, Directed self assembly (DSA) compliant flow with immersion lithography: from material to design and patterning, in *Proceedings for SPIE Advanced Lithography* (2016), pp. 1–11

Chapter 9
Cut Optimization

Line-end cut process is used to create very fine metal wires in sub-14 nm technology. Cut patterns split regularly spaced line patterns into a number of wire segments, some of which become actual routing wires while the remainders are regarded as dummy. In sub-10 nm technology, cuts are smaller than optical resolution limit and a directed self-assembly lithography with multiple patterning (MP-DSAL) is a good candidate for their patterning. Cut optimization for MP-DSAL is addressed in this chapter. The optimization goal is to determine cut locations in such a way that cuts are grouped into manufacturable GPs, which are then assigned to one of the masks without MP coloring conflicts; minimizing wire extensions is also pursued in the process. The optimization problem is formulated as ILP and a fast heuristic algorithm is also presented.

9.1 Introduction

As technology node scales down to sub-7 nm, printing small features becomes challenging task for traditional optical lithography. MP-DSAL is considered as a viable technique for printing contact and via patterns [1–3]; its application for printing line-end cuts (or cuts for short) has also been studied [4, 5].

In line-end cut process, regularly spaced line patterns are first created on a wafer through self-aligned double patterning (SADP) [10]; they are then split into multiple wire segments by cuts. Some segments are used as actual routing wires while the remainders become dummies and serve no function. Process to create cut patterns with MP-DSAL is as follows: cuts that are physically close are clustered (Fig. 9.1a) and a contour that surrounds each cluster, called a guide pattern (GP) image, is synthesized (Fig. 9.1b). If a distance between two GP images is smaller than optical resolution limit, they are patterned on different masks (Fig. 9.1c) and actual GPs are

S. Shim and Y. Shin, *Physical Design and Mask Synthesis for Directed Self-Assembly Lithography*, NanoScience and Technology,
https://doi.org/10.1007/978-3-319-76294-4_9

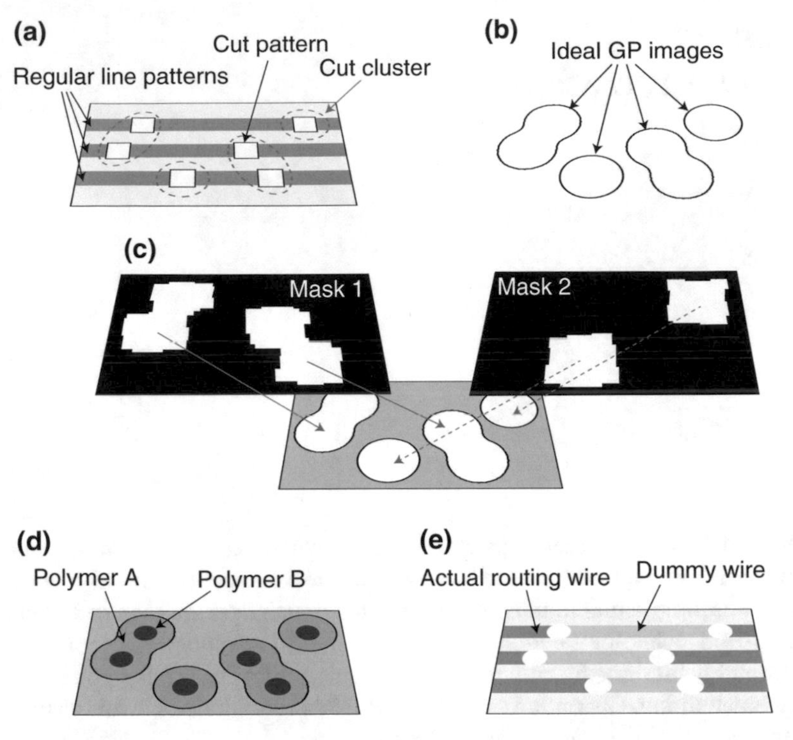

Fig. 9.1 Line-end cut process with MP-DSAL: **a** regular line patterns with cut clusters, **b** ideal GP images of cut clusters, **c** multi-patterning to create GPs on a wafer, **d** GPs filled with BCP, and **e** routing and dummy wires created after polymer B is etched away

created on a wafer by processing each mask one by one. GPs are filled with block copolymers (BCPs), which align by themselves due to forces between BCPs and GP walls during heating and annealing process (Fig. 9.1d). After one type of polymer (polymer B) is etched down to the underlying metal layer, cuts are formed and regular line patterns are split into routing and dummy wires (Fig. 9.1e).

Clustering nearby cuts is not arbitrary. A large and complex GP image is hard to manufacture during DSA process [11] as we have discussed in Chap. 2, so GP can contain only a limited number of cuts and its shape should be simple. In addition, GPs have to be assigned to one of the masks without coloring conflicts. This makes cut clustering problem with mask assignment very difficult. In fact, a similar version of problem in contact and via layers has been shown to be NP-complete [1, 12].

Unlike contacts and vias whose locations are fixed after layout design, cuts can be relocated to avoid forming unmanufacturable clusters and causing coloring conflicts. However, relocating cuts results in wire extensions as shown in Fig. 9.2a. This may negatively affect circuit timing due to increased capacitance. The goal of cut optimization for MP-DSAL, the topic of this chapter, is to determine cut locations together with cut clustering and mask assignment in such a way that all GPs

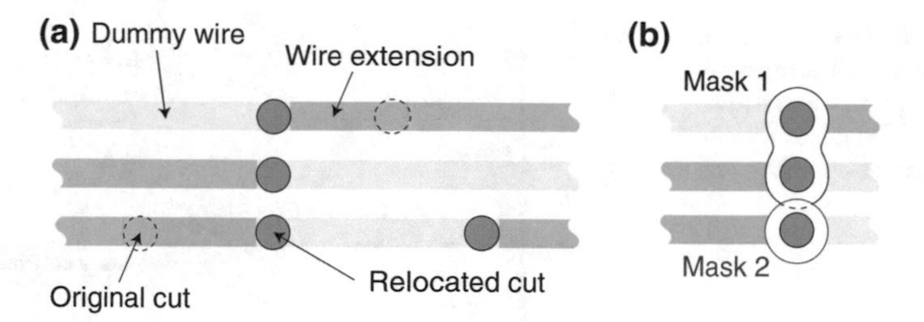

Fig. 9.2 **a** Wire extensions due to relocated cuts, and **b** coloring conflict between clusters assigned to different masks

are manufacturable with smallest coloring conflicts and minimum wire extensions. Small number of coloring conflicts can manually be fixed by designers, so it is aimed to minimized conflicts but not to search for conflict-free solution.

The remainder of this chapter is organized as follows. Section 9.2 briefly discusses critical cut distance in MP-DSAL, and the impact of wire extensions on circuit timing. Section 9.3 presents the ILP formulation along with a heuristic algorithm. Cut optimization algorithms are demonstrated and compared with the existing approach in Sect. 9.4. The chapter is summarized in Sect. 9.5.

9.2 Preliminaries

9.2.1 Critical Cut Distances in MP-DSAL

Suppose p denotes a center-to-center distance or pitch between two cuts; minimum pitch supported by DSAL is denoted by p^-_{dsa}; maximum length that BCP can stretch is denoted by p^+_{dsa}. For any two cuts to be clustered together, their pitch has to be at least p^-_{dsa} but no more than p^+_{dsa}; this range of p is marked as DSA in Fig. 4.2 (see Chap. 4). If their pitch is larger than p_{mp}, which is a minimum pitch supported by multi-patterning, cuts can be assigned to different clusters whose GPs are patterned on different masks. This range of p is denoted as MP in Fig. 4.2. Hence, if a pitch between two cuts lies in $p_{mp} \leq p \leq p^+_{dsa}$ (marked as DSA/MP), they can either be assigned to the same cluster or be assigned to different clusters whose GPs are patterned on different masks. Finally, if a pitch between two cuts is larger than p_{litho}, which is a minimum pitch supported by optical lithography, cuts can freely be assigned to any different clusters. This range of p is marked as single mask in Fig. 4.2.

The values of p^-_{dsa} and p^+_{dsa} are determined by the length of BCP [8] that is employed. The values of p_{mp} and p_{litho} are determined from the BCP length as well as optical resolution limit, which in turn is affected by lithography settings such as wavelength of light source, NA, and illumination shape.

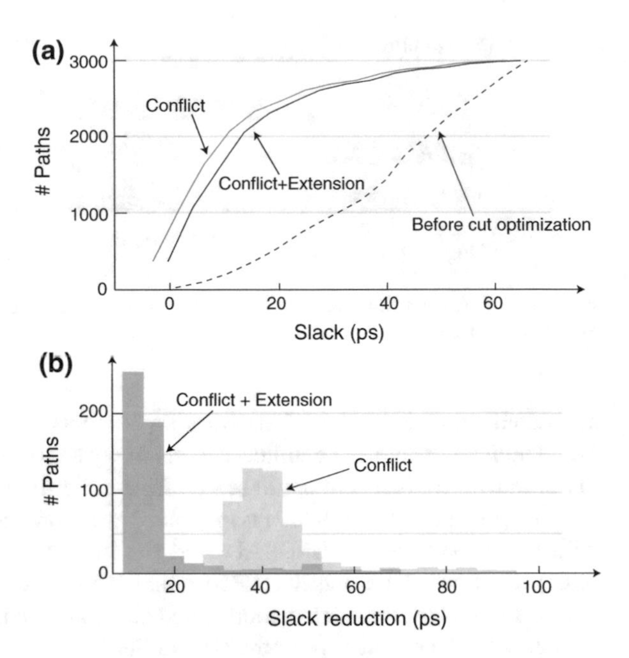

Fig. 9.3 **a** Slack histograms of test circuit spi before and after cut optimization and **b** reduction of slack in critical paths after cut optimization

9.2.2 Wire Extension: Impact on Circuit Timing

Wire extension has been considered as a popular cost function in many studies of cut optimization [4, 19–21]. Its impact on circuit timing is studied in quantitative fashion using 28 nm commercial library.

Two solutions of cut optimization problem are obtained using ILP, whose details are presented in Sect. 9.3. In the first solution denoted by Conflict+Extension in Fig. 9.3, both the number of coloring conflicts and the amount of wire extensions are minimized; this is the approach addressed in this section. The second solution denoted by Conflict serves as a reference; only the number of conflicts is minimized while wire extensions are simply kept below a certain value.

A test circuit spi is used for experiment; its slack histogram before cut optimization is shown in Fig. 9.3a. Slack histogram is shifted to the left (i.e., more paths with smaller slack) in both solutions due to wire extensions. But that of Conflict+Extension is shifted less and exhibits better timing. In Fig. 9.3b, we only take critical paths whose slack is smaller than 50 ps, and the reduction of slack in both solutions is identified with the number of paths in y-axis. Clearly, Conflict+Extension causes less reduction: on average of 15 ps as opposed to 42 ps in Conflict. Figure 9.3 demonstrates that wire extensions should really be minimized in cut optimization problem.

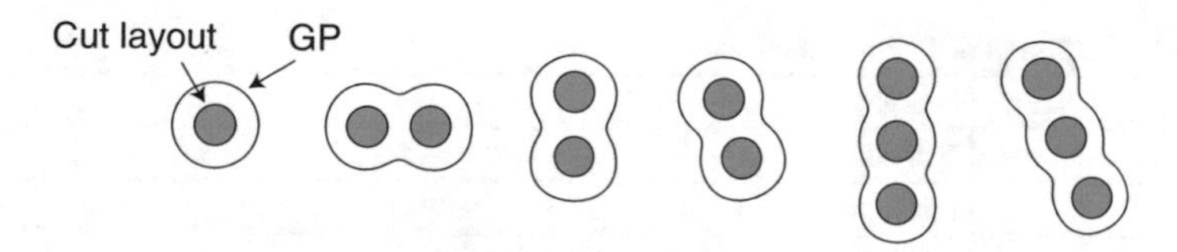

Fig. 9.4 Manufacturable GPs

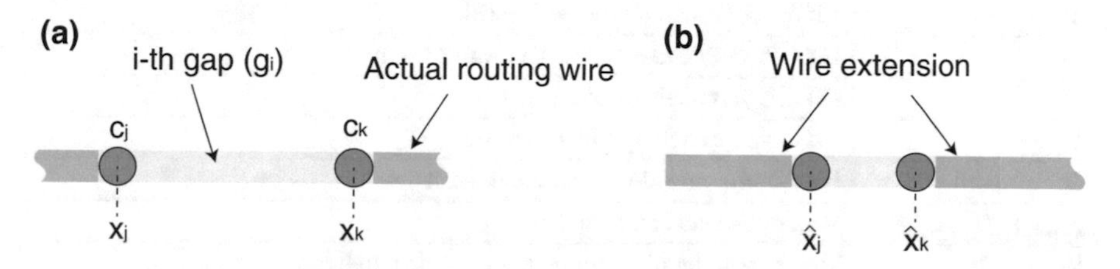

Fig. 9.5 A gap and its corresponding cuts on the left and right **a** before and **b** after the cuts are relocated

9.3 MP-DSAL Cut Optimization

Cut optimization for MP-DSAL determines the location of all cuts while each cut is assigned to a cluster (for DSAL) which is then assigned to a mask (for MP). It guarantees that (1) circuit connectivity remains the same, (2) total wire extensions are minimized, and (3) MP coloring conflicts are kept at a minimum. ILP formulation is first tried to solve this problem, which is then followed by description of heuristic algorithm. It is assumed that only GPs with zero defect probability (i.e., the ones with linearly aligned cuts as illustrated in Fig. 9.4) are allowed.

9.3.1 ILP Formulation

A gap, denoted by g_i, is defined as a space between adjacent actual routing wires on the same metal track as shown in Fig. 9.5a. It is associated with two cuts, left cut (c_j) at position (x_j, y_j) and right cut (c_k) at position (x_k, y_k). Initial location of cuts is assumed to be closest to vias within design rules are violated, i.e., x_j is the smallest and x_k is the largest. It is also assumed, without loss of generality, that wires run in horizontal direction. After cut optimization of c_j and c_k, their new x-coordinates are denoted by $\widehat{x}_j$ and $\widehat{x}_k$, respectively, as shown in Fig. 9.5b. This yields the wire extension of $\widehat{x}_j - x_j$ plus $x_k - \widehat{x}_k$

Let GP_{ik} be a Boolean variable, which is set to 1 if cut c_i assigned to k-th (manufacturable) GP. Another Boolean variable M_{ik} is introduced; it is 1 if c_i's GP resides on k-th mask. Notations of ILP formulation are summarized in Table 9.1. Cut optimization for MP-DSAL is to determine the value of $\widehat{x}_i$, GP_{ik}, and M_{ik} with objective of minimizing total wire extensions and MP coloring conflicts:

Table 9.1 Notations of ILP formulation

g_i	i-th gap
c_i	i-th cut
(x_i, y_i)	Initial location of c_i
$\widehat{x_i}$	x location of c_i after cut optimization
GP_{ik}	1 if c_i is assigned to k-th GP
M_{ik}	1 if c_i' GP is assigned to k-th mask
C_{ij}	1 if GPs of c_i and c_j cause a coloring conflict
O_{ij}	1 if $(\widehat{x_i}, y_i)$ equals to $(\widehat{x_j}, y_j)$
S_{ij}	1 if c_i and c_j belong to the same GP
M_{ij}	1 if c_i and c_j reside on the same mask
$p_{min}(i, j)$	Min. pitch between c_i and c_j
$h^-_{dsa}(i, j)$	Min. horizontal pitch for c_i and c_j to be clustered
$h^+_{dsa}(i, j)$	Max. horizontal pitch for c_i and c_j to be clustered
$h_{mp}(i, j)$	Min. horizontal pitch for c_i and c_j to be assigned to different clusters on different masks
$h_{litho}(i, j)$	Min. horizontal pitch for c_i and c_j to be assigned to different clusters on the same mask
M_{ijk}	1 if k-th bit of mask index of c_i and c_j is same
m_{ik}	k-th bit of mask index of c_i

$$\text{Maximize} \sum (\widehat{x_j} - x_j + x_k - \widehat{x_k}) + \alpha\Big(\sum C_{ij}\Big), \tag{9.1}$$

where α is a constant weight and C_{ij} is equal to 1 if GPs of c_i and c_j cause coloring conflict. The problem is subject to

$$\sum_k GP_{ik} = 1, \quad \forall c_i \tag{9.2}$$

$$x_j \le \widehat{x_j} \le \widehat{x_k} \le x_k, \quad \text{for } c_j \text{ and } c_k \text{, which are left and right cuts of } g_i \text{, respectively} \tag{9.3}$$

$$O_{ij} \le S_{ij} \le M_{ij}, \quad \forall c_i \text{ and } c_j \ (i \ne j) \tag{9.4}$$

for all c_i and c_j such that $\widehat{x_i} \le \widehat{x_j}$ and $p_{min}(i, j) \le p^+_{dsa}$,

$$\widehat{x_j} - \widehat{x_i} \ge h^-_{dsa}(i, j) - \alpha\big[(1 - S_{ij}) + O_{ij} + C_{ij}\big], \tag{9.5}$$

$$\widehat{x_j} - \widehat{x_i} \le h^+_{dsa}(i, j) + \alpha\big[(1 - S_{ij}) + O_{ij} + C_{ij}\big], \tag{9.6}$$

for all c_i and c_j such that $\widehat{x_i} \le \widehat{x_j}$ and $p_{min}(i, j) < p_{mp}$,

$$\widehat{x_j} - \widehat{x_i} \geq h_{mp}(i, j) - \alpha\big[S_{ij} + M_{ij} + C_{ij}\big], \tag{9.7}$$

for all c_i and c_j such that $\widehat{x_i} \leq \widehat{x_j}$ and $p_{min}(i, j) \leq p_{litho}$,

$$\widehat{x_j} - \widehat{x_i} \geq h_{litho}(i, j) - \alpha\big[S_{ij} + (1 - M_{ij}) + C_{ij}\big], \tag{9.8}$$

for all c_i and c_j, $i \neq j$,

$$M_{ijk} \geq 1 - m_{ik} - m_{jk}, \quad k = 1, 2 \tag{9.9a}$$
$$M_{ijk} \leq 1 - m_{ik} + m_{jk}, \quad k = 1, 2 \tag{9.9b}$$
$$M_{ijk} \leq 1 + m_{ik} - m_{jk}, \quad k = 1, 2 \tag{9.9c}$$
$$M_{ijk} \geq m_{ik} + m_{jk} - 1, \quad k = 1, 2 \tag{9.9d}$$

$$2M_{ij} \leq M_{ij1} + M_{ij2}, \tag{9.10a}$$
$$M_{ij} \geq M_{ij1} + M_{ij2} - 1. \tag{9.10b}$$

Inequality (9.2) restricts that one cut is assigned to only one GP. Inequality (9.3) specifies feasible horizontal cut location so that circuit connectivity remains the same. Inequality (9.4) guarantees that two cuts at the same location have to be clustered together, and two cuts in the same cluster have to be assigned to the same mask. Inequalities (9.5)–(9.6) restrict that two cuts can be clustered together if their pitch is within the range of DSA (see Fig. 4.2). $h^-_{dsa}(i, j)$ is obtained by $\sqrt{(p^-_{dsa})^2 - (y_i - y_j)^2}$, and $h^+_{dsa}(i, j)$ is obtained similarly. For two cuts that belong to different clusters whose GPs are patterned on different masks, inequality (9.7) restricts that their pitch has to be larger than or equal to p_{mp}. If cuts are assigned to clusters whose GPs are patterned on the same mask, inequality (9.8) ensures that their pitch is larger than or equal to p_{litho}. Note that $h_{mp}(i, j)$ and $h_{litho}(i, j)$ are obtained in similar manner to $h^-_{dsa}(i, j)$. A set of inequalities (9.9a)–(9.9d) is equivalent to XNOR operation (e.g., $M_{ij1} = m_{i1}\text{XNOR}m_{j1}$), which sets M_{ijk} to 1 if the k-th bit of mask index of i-th and j-th cut are identical (i.e., $m_{ik} = m_{jk}$). Inequalities (9.10a)–(9.10b) correspond to AND operations (i.e., $M_{ij} = M_{ij1}$ AND M_{ij2}), which determines M_{ij}.

In this sample ILP formulation, four masks are assumed. In case of double and triple patterning, additional constraints $m_{i1} = 0$ and $m_{i1} + m_{i2} \leq 1$ have to be added, respectively.

9.3.2 *Heuristic Algorithm*

A fast heuristic algorithm is presented, which uses divide-and-conquer strategy to solve the cut optimization problem for large circuits. A conflict graph is first

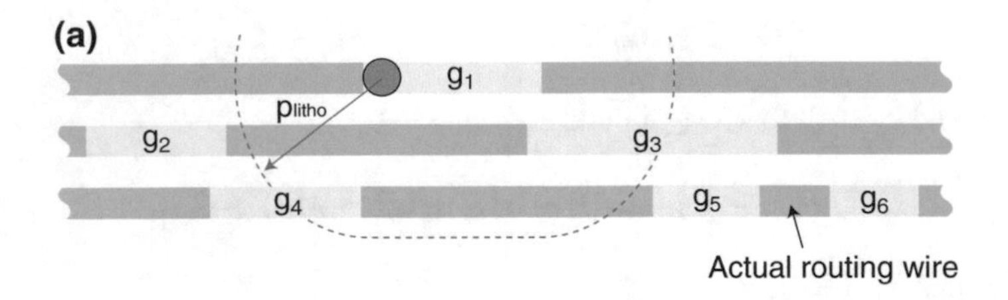

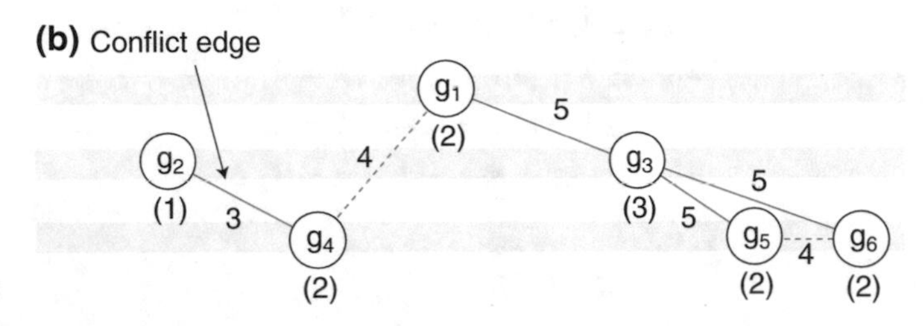

Fig. 9.6 **a** Example metal layout, and **b** its corresponding conflict graph

constructed to represent all possible coloring conflicts on a given layout. By ignoring some edges, the graph is partitioned into many small trees, which can independently be solved to an optimal. Solutions from all trees are merged, and any remaining coloring conflict between trees is removed by local adjustment.

Conflict Graph Construction: For a given layout, possible coloring conflict between gaps is identified by a feasible location of their cuts after optimization. If two gaps are closer than a minimum distance p_{litho} (see g_1 and g_4 in Fig. 9.6a), coloring conflict may exist between them. By representing each gap as a vertex, an edge is drawn to represent possible coloring conflict as shown in Fig. 9.6b.

Each vertex is assigned with a weight equal to the number of connected edges (see the numbers in brackets). A vertex with larger weight is more likely to have coloring conflict. Such vertex and its neighbors should be included in the same tree after partition. For this purpose, an edge has a weight equal to the total weight of all its connected vertices. Partitioning is done by initializing all vertices to be trees and sequentially connecting them with an edge in a decreasing weight order. An edge is excluded if it causes a cycle, or if it causes tree that is too big. If a tree is limited to size 4, the conflict graph in Fig. 9.6a is divided into two trees (see solid edges in Fig. 9.6b).

Solving Tree to an Optimal: For each tree, a matching graph is constructed as shown in Fig. 9.7a. A vertex of the graph represents a feasible configuration of cuts on each gap. Example vertices on gap $g1$ and $g3$ are shown in Fig. 9.7b. There exists an edge connecting two vertices if they belong to two connected gaps of a tree and form a valid cut configuration. For example, v_1 is connected to v_4 because those vertices have matched GPs on the left. v_1, however, is not connected to v_5 because their left

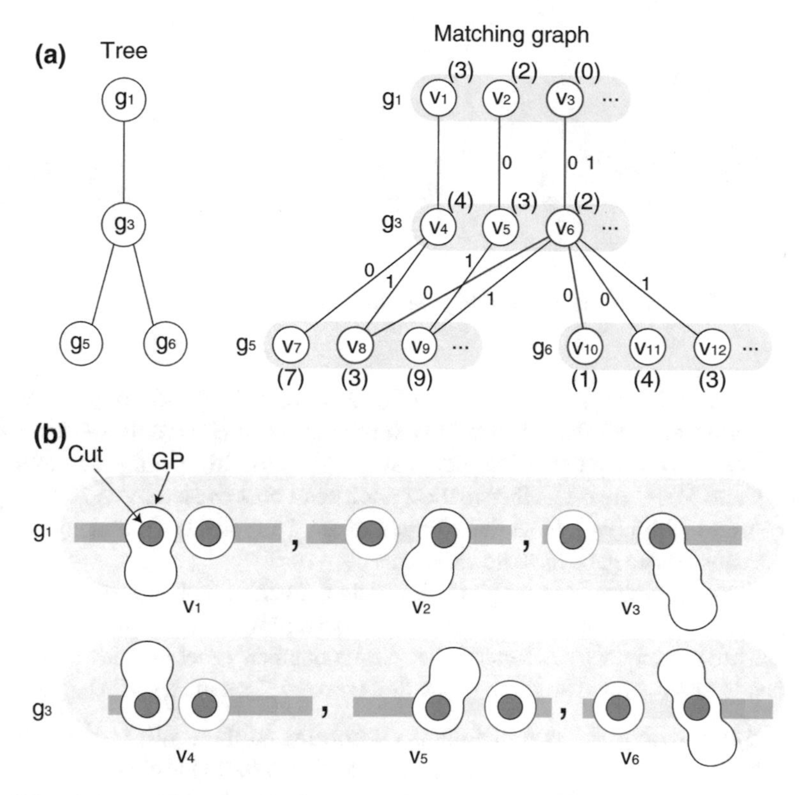

Fig. 9.7 **a** A tree and its corresponding matching graph, and **b** examples of vertices from gap $g1$ and $g3$

GPs are incompatible. The same concept is applied for configuration with three-cut GPs (see v_3 and v_6).

Each vertex is associated with a weight value that is equal to amount of wire extensions (see number in brackets). If two matching vertices have a coloring conflict, corresponding edge is assigned with a weight of 1 (see right GP of v_1 and v_4); otherwise, edge is assigned with a weight of 0 (see v_2 and v_5). Since the goal is to find cuts configuration for all gaps with a minimum cost defined in (9.1), one gap of tree is randomly picked as a root say g_1, and a tree is traversed in post-order ($g_5 \rightarrow g_6 \rightarrow g_3 \rightarrow g_1$). While visiting each gap, cost at its corresponding vertices is calculated as follows:

$$COST(v_i) = w(v_i) + \sum_{G_i} \min_j \big[W \cdot w(e_{ij}) + COST(v_j)\big], \tag{9.11}$$

where $w(v_i)$ is a weight of vertex v_i, G_i denotes children of a gap that vertex v_i represents, and j is feasible configuration of a gap in G_i. $w(e_{ij})$ is a weight of an edge between v_i and v_j, and W is the large constant used in (9.1).

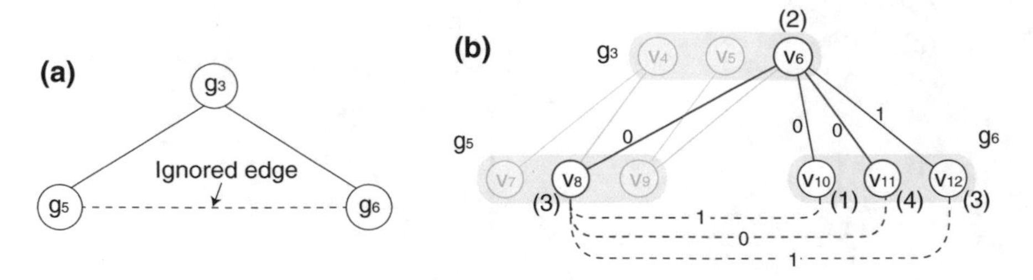

Fig. 9.8 **a** An ignored edge, and **b** matching graph for local adjustment

Cost computation at v_6 is illustrated in Fig. 9.7a. v_6 corresponds to gap g_3 which has two children, g_5 and g_6. A cost at v_6 turns out to be 6 ($= 2 + 3 + 1$): 2 is a weight of v_6; 3 is the smallest additional cost out of all feasible configurations of g_5, 0+3 (v_8), and W+9 (v_9); 1 is the smallest additional cost caused by g_6, 0+1 (v_{10}), 0+4 (v_{11}), and W+3 (v_{12}). Note that a vertex of any leaf gaps has a cost equal to its weight because those gaps have no children.

Once a cost at vertices of root gap is computed, a vertex with the smallest cost is selected, and its descendant vertices that are associated with the minimum cost configuration are subsequently selected in top-down manner. A set of selected vertices corresponds to an optimal solution of the tree (see red lines in Fig. 9.7a).

Local Adjustment: After optimizing cuts according to the result in the previous step, each edge that has been dropped during graph partitioning is checked to see if it causes a coloring conflict. If a conflict is observed, it is resolved by local adjustment. An edge of larger weight is considered first since it is likely to have more conflicts and its conflict is hard to resolve in latter iteration.

Local adjustment is performed by finding an alternative configuration of cuts on two gaps with conflict, while other cuts are assumed to be fixed. Suppose that v_6, v_8, and v_{10} are selected as a solution in the previous step. When considering a dropped edge $g_5 - g_6$ as shown in Fig. 9.8a, there exists a coloring conflict. This is resolved by finding a new configuration of cuts on g_5 and g_6 with the smallest sum of edge and vertex weight. New configuration is then $v_6 - v_8 - v_{11}$ with total weight equal to 9 (see red lines in Fig. 9.8b). This number is smaller than total weight of $v_6 - v_8 - v_{10}(6+W)$ or $v_6 - v_8 - v_{12}(8 + 2W)$.

9.4 Experiments

MP-DSAL cut optimization in Sect. 9.3 is implemented with C++, and GUROBI [22] is used as an ILP solver. Test circuits are taken from Open Cores [23] and ITC99 benchmark [24]; they are synthesized using 28-nm industrial library. M2 layout is taken and is appropriately shrunk to follow virtual sub-7 nm design rule which is assumed. Metal track pitch is 35 nm, and cut size is assumed to be 18 nm by 18 nm.

Table 9.2 Comparison of three-cut optimization approaches

Circuits	# Cuts	Approach in [4]		ILP			Heuristic		
		# Conflicts	ΔWL (μm)	# Conflicts	ΔWL (%)	Runtime (m)	# Conflicts	ΔWL (%)	Runtime (m)
b12	4,776	2,122	518	0	4	14	2	11	1
spi	10,658	4,340	1,217	0	7	27	3	9	1
des3_area	16,164	8,425	1,685	0	9	105	7	12	1
b14	23,220	10,673	2,532	0	8	134	12	10	2
b21	46,502	20,925	5,702	0	7	522	17	8	3
aes_core	79,132	45,044	7,620	–	–	–	50	13	5
b18	121,202	51,977	15,882	–	–	–	40	8	8
ethernet	280,379	109,519	44,898	–	–	–	108	8	21
des3_perf	360,624	170,937	51,332	–	–	–	179	8	24
Average		47,107	14,504	0	7	161	46	9	8

The number of cuts of test circuits ranges from 5 k (b12) to 360 k (des3_perf) as shown in Table 9.2. Double patterning with ArF immersion lithography (1.35 NA) is assumed. Critical cut distances in MP-DSAL are assumed as follows: $p_{dsa}^{-} = 30$ nm, $p_{dsa}^{+} = 45$ nm, $p_{mp} = 39$ nm, and $p_{litho} = 87$ nm.

A simple-minded approach proposed in [4] is also implemented for comparison. It performs cut optimization problem for basic DSAL (without MP) for cuts on every three consecutive metal tracks, and implicitly assigns GPs on those tracks to one mask; this is repeated for cuts in the next three consecutive metal tracks with other mask. When applying this approach, many coloring conflicts remain because the conflicts between clusters assigned to different masks are ignored. ILP completed the cut optimization for five small circuits (see column 5–7). Results show that ILP can remove all coloring conflicts and reduce total wire extensions by 93% on average compared to the approach [4] being applied. The heuristic algorithm achieves a similar result with less than 1% coloring conflict and 91% reduction in total wire extensions (see column 8–10).

Two masks (double patterning or DP) are assumed in Table 9.2. With three masks (triple patterning or TP), wider solution space can be explored. This makes coloring conflict easier to solve and reduces total wire extensions. ILP and heuristic algorithms for TP are applied to all test circuits in Table 9.2. Average number of coloring conflicts and total wire extensions are compared in Fig. 9.9; the results of cut optimization for TP are normalized to those for DP. In TP, no coloring conflict remains after ILP and heuristic are applied. Total wire extensions are reduced by 50% with ILP. Heuristic achieves similar result with 35% reduction in total wire extensions.

On M2 layer that is used, power rails run above and below every cell row. Since the rail is thick enough to prevent any coloring conflict between clusters across it, MP-DSAL cut optimization algorithm can be applied to cuts on each cell row independently. As a cell row becomes longer, more cuts are included and more time are required for optimization. This is demonstrated by performing ILP for DP on test circuit (b12) with varying aspect ratio of placement region. As shown in Fig. 9.10,

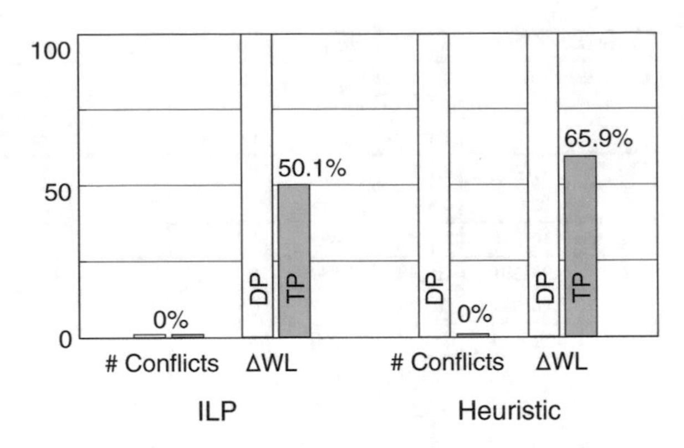

Fig. 9.9 Average number of coloring conflicts and total wire extensions when optimizing cuts for TP and DP with ILP and heuristic algorithm

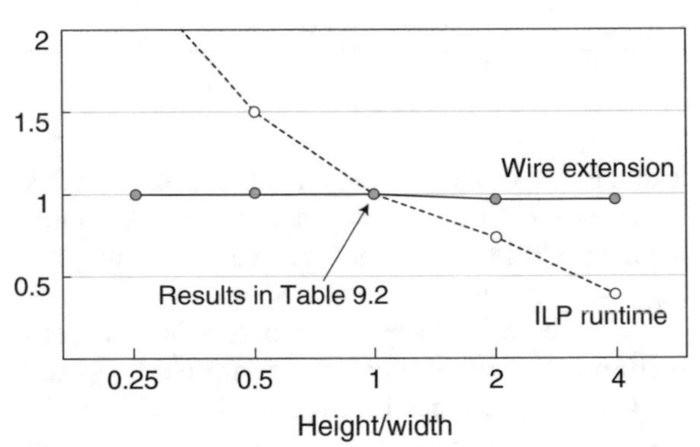

Fig. 9.10 Total wire extensions and runtime of cut optimization on test circuit b12 with different aspect ratios of placement region

runtime increases noticeably when placement width increases. Total wire extensions are, however, almost the same with the result shown in Table 9.2. This is because total number of cuts on placement region slightly changes with the aspect ratio, and likeliness to have coloring conflict is almost the same.

9.5 Conclusion

A cut optimization problem for MP-DSAL has been addressed. ILP for the problem has been formulated, and fast heuristic algorithm has also been presented. The heuristic employs divide-and-conquer strategy. A conflict graph corresponding to a given layout is constructed and partitioned into a number of trees; each tree is solved to an optimal. Solution of all trees is combined together, and local adjustment is performed to remove any coloring conflicts that occur between the solutions. Demonstrations with sub-7 nm technology have indicated that the heuristic algorithm achieves a comparable result to ILP; it yields a solution with less than 1% coloring conflicts and 91% reduction in total wire extensions compared to a simple existing approach.

References

1. Y. Badr, A. Torres, P. Gupta, Mask assignment and synthesis of DSA-MP hybrid lithography for sub-7nm contacts/vias, in *Proceedings of the Design Automation Conference* (2015), pp. 70:1–70:6
2. J. Ou, B. Yu, D. Pan, Concurrent guiding template assignment and redundant via insertion for DSA-MP hybrid lithography, in *Proceedings of the International Symposium on Physical Design* (2016), pp. 39–46
3. S. Shim, W. Chung, Y. Shin, Redundant via insertion for multiple-patterning directed-self-assembly lithography, in *Proceedings of the Design Automation Conference* (2016), pp. 41:1–41:6
4. Z.W. Lin, Y.W. Chang, Double-patterning aware DSA template guided cut redistribution for advanced 1-D gridded designs, in *Proceedings of the International Symposium on Physical Design* (2016), pp. 47–54
5. W. Ponghiran, S. Shim, Y. Shin, Cut mask optimization for multi-patterning directed self-assembly lithography, in *Proceedings of the Design, Automation and Test in Europe (DATE)* (2017), pp. 1498–1503
6. M. Muramatsu, M. Iwashita, T. Kitano, T. Toshima, M. Somervell, Y. Seino, D. Kawamura, M. Kanno, K. Kobayashi, T. Azuma, Nanopatterning of diblock copolymer directed self-assembly lithography with wet development. J. Micro/Nanolithography, MEMS, MOEMS **11**(3), 1–6 (2012)
7. Y. Seino, H. Yonemitsu, H. Sato, M. Kanno, H. Kato, K. Kobayashi, A. Kawanishi, T. Azuma, M. Muramatsu, S. Nagahara, T. Kitano, T. Toshima, Contact hole shrink process using graphoepitaxial directed self-assembly lithography. J. Micro/Nanolithography, MEMS, MOEMS **12**(3), 1–6 (2013)
8. H. Yi, X. Bao, R. Tiberio, P. Wong, Design strategy of small topographical guiding templates for sub-15nm integrated circuits contact hole patterns using block copolymer directed self-assembly, in *Proceedings of the SPIE Advanced Lithography* (2013), pp. 1–9
9. S. Shim, S. Cai, J. Yang, S. Yang, B. Choi, Y. Shin, Verification of directed self-assembly (DSA) guide patterns through machine learning, in *Proceedings of the SPIE Advanced Lithography* (2015), pp. 1–8
10. H. Zhang, Y. Du, M.D.F. Wong, K.Y. Chao, Mask cost reduction with circuit performance consideration for self-aligned double patterning, in *Proceedings of the Asia South Pacific Design Automation Conference* (2011), pp. 787–792
11. S. Shim, W. Chung, Y. Shin, Defect probability of directed self-assembly lithography: fast identification and post-placement optimization, in *Proceedings of the International Conference on Computer Aided Design* (2015), pp. 404–409
12. Z. Xiao, C. Lin, M. Wong, Contact layer decomposition to enable DSA with multi-patterning technique for standard cell based layout, in *Proceedings of the Asia South Pacific Design Automation Conference* (2016), pp. 1–8
13. H. Yi, X. Bao, R. Tiberio, P. Wong, Design strategy of small topographical guiding templates for sub-15nm integrated circuits contact holepatterns using block copolymer directed self-assembly, in *Proceedings of the SPIE Advanced Lithography* (2013), pp. 1–9
14. Z. Xiao, Y. Du, H. Tian, M. Wong, H. Yi, H. Wong, H. Zhang, Directed self-assembly (DSA) template pattern verification, in *Proceedings of the Design Automation Conference* (2014), pp. 1–6
15. H. Yi, L. Azat, P. Wong, Computational simulation of block copolymer directed self-assembly in small topographical guiding templates, in *Proceedings of the SPIE Advanced Lithography* (2013), pp. 1–7
16. L. Azat, G. Grant, P. Moshe, S. Gerard, W. Wong, J. Xu, Y. Zou, Computational simulations and parametric studies for directed self-assembly process development and solution of the inverse directed self-assembly problem. Jpn. J. Appl. Phys. **53**(6S), 1–8 (2014)

17. N. Laachi, K.T. Delaney, B. Kim, S. Hur, R. Bristol, D. Shykind, C.J. Weinheimer, G.H. Fredrickson, Self-consistent field theory investigation of directed self-assembly in cylindrical confinement. J. Polym. Sci. Part B, Polym. Phys. **53**(2), 142–153 (2015)
18. M.C. Smayling, V. Axelrad, 32nm and below logic patterning using optimized illumination and double patterning, in *Proceedings of the SPIE Advanced Lithography* (2009), pp. 1–10
19. Z. Xiao, Y. Du, M. Wong, H. Zhang, DSA template mask determination and cut redistribution for advanced 1D gridded design, in *Proceedings of the SPIE Advanced Lithography* (2013), pp. 1–8
20. J. Ou, B. Yu, J.R. Gao, D. Pan, M. Preil, A. Latypov, Directed self-assembly based cut mask optimization for unidirectional design, in *Proceedings of the Great Lakes Symposium on VLSI* (2015), pp. 83–86
21. Z.W. Lin, Y.W. Chang, Cut redistribution with directed self-assembly templates for advanced 1-D gridded layouts, in *Proceedings of the Asia South Pacific Design Automation Conference* (2016), pp. 89–94
22. Gurobi Optimization, Inc., Gurobi optimizer reference manual, http://www.gurobi.com/
23. Opencores, http://www.opencores.org/
24. ITC99, http://www.cerc.utexas.edu/itc99-benchmarks/

Chapter 10
Summary of The Book

In DSAL, contacts (or vias) are indirectly formed through guide patterns (GPs). Thus, the integrity of GP is very important to obtain desirable contacts on a wafer. Since GP is created by traditional lithography, it may have some errors when its shape is large and complex [1, 2], which affect final contact patterns. Such limitation of GP shape calls for careful considerations in physical design. It has been also argued that conventional mask synthesis and verification for traditional lithography are obsolete in DSAL. In this context, this book has addressed problems on physical design and mask synthesis as follows (Fig. 10.1).

- **Physical Design**
  - **Placement optimization**: Conventional placement method generates many undesirable intercell clusters. They can be removed by inserting whitespace in between every problematic cell pairs, but this comes at the cost of large area overhead. Post-placement optimization [2] flips some cells and swaps some cells with adjacent cells so that the number of whitespaces inserted can be minimized; wirelength may increase due to cell displacement. Automatic placer that minimizes both defect probability and total wirelength has also been introduced.
  - **Redundant via insertion**: The goal of redundant via insertion for DSAL [3] is to maximally insert redundant vias for via manufacturability while original and redundant vias do not form undesirable via cluster. This problem has been addressed by employing a conflict graph, which represents all conflicts of redundant via positions due to undesirable cluster and design rules, and solving maximum independent set (MIS) of the conflict graph.
  - **Physical design optimization for MP-DSAL**: In MP-DSAL, contacts and vias are clustered and their GP is assigned to one of the masks (colors). Therefore, the above two problems have been extended considering MP coloring conflict. Post-placement optimization for MP-DSAL compliant contact layout [4] has been addressed. Redundant via insertion for MP-DSAL [5] has been also addressed;

S. Shim and Y. Shin, *Physical Design and Mask Synthesis for Directed Self-Assembly Lithography*, NanoScience and Technology,
https://doi.org/10.1007/978-3-319-76294-4_10

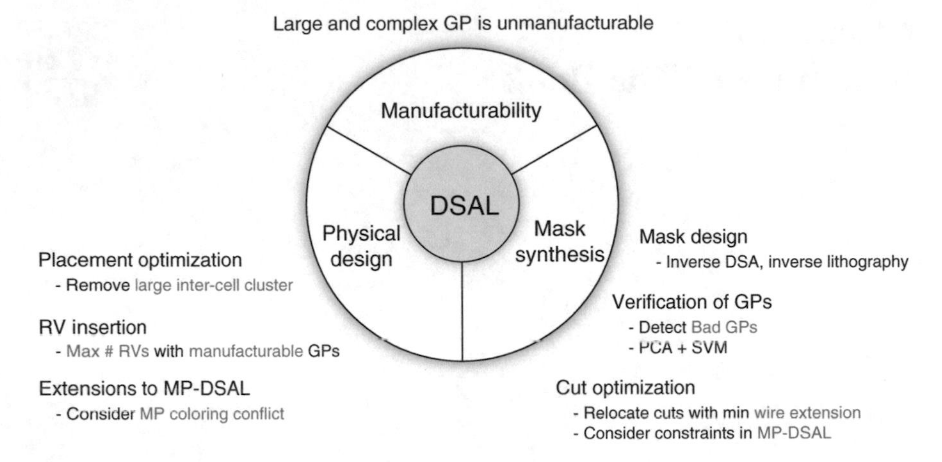

Fig. 10.1 Summary of this book

its goal is to insert maximum number of redundant vias while all via clusters are manufacturable and no MP coloring conflict occurs.

- **Mask synthesis**
  - **Mask design**: Two key problems have been addressed in DSAL mask design [6]: (1) inverse DSA is to find ideal GP image so that desired contact pattern can be produced after DSA process; (2) inverse lithography is to find perfect mask image so that the ideal GP image can be produced after lithography process. Inverse DSA and inverse lithography have been extended to handle process variations.
  - **Verification of GPs**: GP synthesis cannot be perfect simply by construction, so its verification is necessary in DSAL. It was shown that functions for the verification can be constructed and be reliably used [7, 8]. The verification function is constructed in three key steps as follows: (1) necessary yet a small number of test GPs are prepared; (2) each GP is represented as a small number of geometric parameters, possibly applying PCA to even reduce this number; (3) a hyperplane as a verification function is determined in the parameter space, the verification function is constructed using a support vector machine.
  - **Cut optimization**: The goal of cut optimization is to determine locations of cuts, cut clusters, and mask assignment [9] so that all clusters are manufacturable, MP coloring conflicts do not occur, and wire extension can be minimized. This problem has been formulated as ILP, and more practical heuristic algorithm has also been presented.

References

1. S. Shim, Y. Shin, Physical design and mask optimization for directed self-assembly lithography (DSAL), in *Proceedings of the International Conference on Very Large Scale Integration (VLSI-SoC)* (2015), pp. 80–85
2. S. Shim, W. Chung, Y. Shin, Defect probability of directed self-assembly lithography: fast identification and post-placement optimization, in *Proceedings of the International Conference on Computer Aided Design* (2015), pp. 404–409
3. W. Chung, S. Shim, Y. Shin, Redundant via insertion in directed self-assembly lithography, in *Proceedings of the Design, Automation and Test in Europe Conference and Exhibition* (2016), pp. 55–60
4. S. Shim, W. Chung, Y. Shin, Placement Optimization for MP-DSAL Compliant Layout, in *Proceedings of the International Conference on IC Design and Technology (ICICDT)* (2016), pp. 1–4
5. S. Shim, W. Chung, Y. Shin, Redundant via insertion for multiple-patterning directed-self-assembly lithography, in *Proceedings of the Design Automation Conf.* (2016), pp. 41:1–41:6
6. S. Shim, Y. Shin, Mask optimization for directed self-assembly lithography: inverse DSA and inverse lithography, in *Proceedings of the Asia South Pacific Design Automation Conference* (2016), pp. 83–88
7. S. Shim, S. Cai, J. Yang, S. Yang, B. Choi, Y. Shin, Verification of directed self-assembly (DSA) guide patterns through machine learning, in *Proceedings of the SPIE Advanced Lithography* (2015), pp. 1–8
8. S. Shim, Y. Shin, Fast verification of guide patterns for directed self-assembly lithography, in *IEEE Transactions on CAD of Integrated Circuits and Systems*, to be published
9. W. Ponghiran, S. Shim, Y. Shin, Cut mask optimization for multi-patterning directed self-assembly lithography, in *Proceedings of the Design, Automation and Test in Europe (DATE)* (2017), pp. 1498–1503

Index

S. Shim and Y. Shin, *Physical Design and Mask Synthesis for Directed Self-Assembly Lithography*, NanoScience and Technology,
https://doi.org/10.1007/978-3-319-76294-4

M

N

O

P

Q

R

S

Printed in the United States
By Bookmasters